W0036570

acoustic imaging

cameras, microscopes, phased arrays,
and holographic systems

acoustic imaging

acoustic imaging

cameras, microscopes, phased arrays,
and holographic systems

Edited by

Glen Wade
University of California at Santa Barbara

SPRINGER SCIENCE+BUSINESS MEDIA, LLC

Library of Congress Cataloging in Publication Data
Main entry under title:

Acoustic imaging

"Lectures on acoustical holography and imaging presented at the University of
California in Santa Barbara, California, March, 1975."
  Includes bibliographical references and index.
  1. Acoustic holography—Congresses. I. Wade, Glen.
TA1550.A26   774   76-21

*Lectures on Acoustical Holography and Imaging presented at the
University of California in Santa Barbara, California, March, 1975*

ISBN 978-1-4757-0826-4   ISBN 978-1-4757-0824-0 (eBook)
DOI 10.1007/978-1-4757-0824-0

© *1976 Springer Science+Business Media New York*
Originally published by Plenum Press, New York in 1976
Softcover reprint of the hardcover 1st edition 1997

All rights reserved

*No part of this book may be reproduced, stored in a retrieval system, or transmitted,
in any form or by any means, electronic, mechanical, photocopying, microfilming,
recording, or otherwise, without written permission from the Publisher*

Preface

The ability to "see" with sound has long been an intriguing concept. Certain animals, such as bats and dolphins, can do it readily, but the human species is not so endowed by nature. However this lack of natural ability can be overcome by applying the appropriate technology. For example, in various laboratories recently, workers have been able to obtain true-focussed, orthographic images in real time of objects irradiated with sound rather than light. Crossectional images have been available for a much longer period of time, stemming from the development of the pulse-echo techniques first used in the sonar systems of World War I. By now a wide variety of system concepts for acoustic imaging exist. Some of the newer systems range from the purely holographic to the purely lens types.

It is apparent that ultrasonic energy can give an image of an object not obtainable with light. For example, a particular object may be embedded in material completely opaque to light but relatively transparent to sound. In addition, soft tissue in the human body frequently provides little contrast for optical radiation, and for x-rays, but large contrast for sonic radiation. Also, the cumulative effect of x-rays may impose severe damage to these tissues, whereas low-intensity sound may be entirely safe.

For these reasons, this comparatively new field, combining as it does in many instances acoustics and optics, is bringing to a number of areas of scientific, medical and engineering endeavor, investigative instruments of great power. The technological developments in this field are already having significant effects not only in diagnostic medicine as implied above, but also in nondestructive testing, seismic sensing, oceanic search and even microscopy.

This book deals with these themes on a level ranging from purely tutorial exposition to treatments describing the latest research progress. Topics are covered from both analytical and experimental viewpoints and include the

general background, the historical development, the categorization and characterics of the modern systems, the
detailed descriptions of these systems, and their applications for acoustic imaging. The book is replete with
examples of the applications, particularly in the medical
area in terms of diagnostic imaging of internal bodily
structure. Before the final draft was completed, the
various presentations in the book were tested on students
in an Extension Course given at the University of California
entitled "Acoustical Holography and Imaging."

The editor gratefully acknowledges the whole-hearted
cooperation of the contributing authors and the institutions
with which they are associated. He also expresses appreciation for the advice and assistence of the staff at the
University of California Extension, in particular Dr. Sidney
Goren, Ms. Judy Weisman, Mr. John Maxwell and Mr. Larry
Nicklin.

Glen Wade

Santa Barbara, California
March 1976

Contents

CONTENTS

Chapter 1

THE PROPOGATION OF ACOUSTIC WAVES

Byron B. Brenden

Holosonics, Inc.

2950 George Washington Way, Richland, Wa   99352

1.1   INTRODUCTION

The idea of imaging using acoustical energy is intriguing in itself aside from concern for its practical application. It may be compared to imaging in the infrared,
ultraviolet, microwave or x-ray portions of the electromagnetic spectrum but using acoustical energy for imaging
is quite different in the sense that it involves more than
an extension of vision into the far reaches of the electromagnetic spectrum. In that sense, acoustical imaging has
its own distinct characteristics and takes its place with
electron optics as an extension of vision into non-electro-
magnetic realms. In the final stage, however, all forms
of imaging must incorporate a step which returns to the
realm of light to produce the final visual impression.

Although acoustical imaging is a uniquely different
form of imaging, many of the elements of imaging carry over
from the optical imaging art. It is, in fact, difficult to
avoid the use of the terms "optics" or "optical" when discussing acoustical imaging system since different but
equivalent terms have not come into common usage. Instead
of saying "acoustical lens system" one is tempted to refer
to the "optics" of the acoustical imaging system. The
term "optical axis" has a meaning which the term "acoustical
axis" does not seem to convey. The lenses and mirrors of
optics have quite recognizable counterparts in acoustical
imaging systems and, although the materials of which they are
fabricated and the methods of fabrication are quite different, many of the design equations are the same. Snell's

1

Law of optics has its exact counterpart in acoustical lens design. The design problem in acoustics is, however, slightly more complicated than the similar problem in optics because of the existence of two types of waves, namely, longitudinal and shear waves.

This chapter is devoted to a review of the basic characteristics of the propogation of acoustical waves in liquids and solids. With regard to terminology, the terms "acoustic", "sound" and "sonic" are considered to be unrestricted in frequency range. Whereas the term "audio" defines a frequency range from approximately 15Hz to 20KHz over which the human ear is responsive, and the term "ultrasonic" refers to frequencies above the audio range, no such restrictions are placed upon the terms "acoustic", "sound" or "sonic".

1.2   DESCRIPTION OF THE WAVE

1.2.1   Description in Terms of Particle Displacement

An acoustical wave of plane wavefront unbounded in either space or time and propagating in the positive z-direction may be represented by the expression

$$w = w_0 \cos(\Omega t - Kz) \tag{1.1}$$

In this expression w represents the instantaneous magnitude of the displacement of an elemental volume (a particle) of material from its normal or undisturbed position when no wave is present, w_0 represents the maximum value (amplitude) of w, t represents the time elapsed from some arbitrary beginning time and z is the distance along the z-axis of an x,y,z coordinate system.

The quantities Ω and K are defined by the equations

$$K = 2\pi / \Lambda \tag{1.2}$$

and

$$\Omega = 2\pi c / \Lambda \tag{1.3}$$

where Λ is the wavelength and c is the velocity of the acoustical wave.

Equation (1.1) describes equally well the scalar properties of both lontgutidinal and transverse waves.   Transverse acoustical waves will be referred to as shear waves
throughout the remainder of this chapter.

1.2.2 Description In Terms of Pressure

A description of an acoustical wave in terms of pressure can be derived with reference to Fig. 1.1. If we
confine our description to a longitudinal wave unbounded
in space or time and propagating in the positive z-direction,
any given elemental volume, $V = \Delta x \Delta y \Delta z$, will be moving
along a line parallel to the z-axis. We shall refer to the
material within this elemental volume as a particle. The
pressure in any given x-y plane is constant but, due to
the presence of the wave, the pressure varies in z-direction. Because of the variation of pressure within the
volume, there is a net force in z-direction which does two
things. It moves the particle a distance w within a given
time interval dt and it changes the z-dimension from its

initial value, Δz, to a new value $\Delta z (1 + \frac{\partial w}{\partial z})$. The resulting change in volume is

$$\Delta V = V(1 + \frac{\partial w}{\partial z}) - V \qquad (1.4)$$

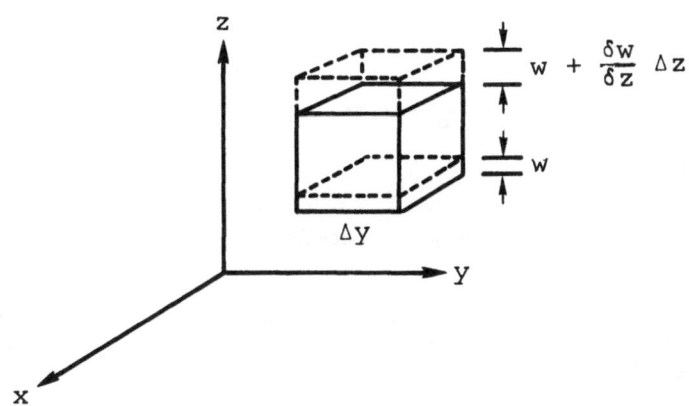

Fig.1.1. Response of an incremental volume to the
pressure within an acoustical wave.

Thus it can be seen that

$$\frac{\Delta V}{V} = \frac{\partial w}{\partial z} \qquad (1.5)$$

One of the characteristics of any material, be it solid, liquid, or gas, is its compressibility. Compressibility is defined as the change in volume per unit volume, $\Delta V/V$, produced by a change in pressure ΔP. If no heat flows during the compression or expansion of the volume the resulting measured value of compressibility

$$\beta = - \frac{\Delta V}{V \Delta P} \qquad (1.6)$$

is known as the adiabatic compressibility. Heat transfer is normally very small during the rapid oscillations of an acoustical wave so adiabatic conditions prevail.

Comparing Eqs. (1.5) and (1.6) and replacing ΔP by p, we obtain

$$p = - \frac{1}{\beta} \frac{\partial w}{\partial z} \qquad (1.7)$$

Upon differentiation of Eq. (1.1) we find that

$$p = - \frac{K}{\beta} w_o \sin(\Omega t - Kz) \qquad (1.8)$$

and setting

$$P_o = - \frac{K}{\beta} w_o \qquad (1.9)$$

we have

$$p = p_o \sin(\Omega t - Kz) \qquad (1.10)$$

Equation (1.9) provides a useful relationship between the particle displacement amplitude w_o and the pressure amplitude of p_o.

Considering again the elemental volume of Fig. 1.1 and the variation of pressure in the z direction we note that if the pressure at the bottom of the particle is P then the

pressure at the top of the particle is $P + \frac{\partial P}{\partial z} \Delta z$ so that
the net force on the particle is

$$F = - \frac{\partial P}{\partial z} \Delta x \Delta y \Delta z \qquad (1.11)$$

and the force per unit volume is

$$F_v = - \frac{\partial P}{\partial z} \qquad (1.12)$$

The pressure P was considered to be the total pressure, but if we consider P to be made up of a constant
positive pressure P_0 and a varying component p which may
be either positive or negative then

$$P = P_o + p \qquad (1.13)$$

and

$$F_v = - \frac{\partial p}{\partial z} \qquad (1.14)$$

Equating the force per unit volume, as given by Eq.
(1.14), to the product of the mass per unit volume, ρ, and

the acceleration of the particle, $\frac{\partial^2 w}{\partial t^2}$, we have

$$- \frac{\partial p}{\partial z} = \rho \frac{\partial^2 w}{\partial t^2} \qquad (1.15)$$

The expressions for p and for w are found in Eqs. (1.1)
and (1.8). If these expressions are substituted into Eq.
(1.15) we find that

$$p_o K = -\rho \Omega^2 w_o \qquad (1.16)$$

from which a second relationship between the pressure amplitude p_0 and the particle displacement amplitude may be
deduced, namely

$$P_o = - \Omega \rho c w_o \qquad (1.17)$$

1.3 Material Properties and Wave Velocity

Although ultrasonic velocities are readily measured
directly it is helpful to know how they depend upon the

physical properties of materials. Equations (1.9) and
(1.16) may be combined to show that

$$\frac{K^2 w_o}{\beta} = \rho \Omega^2 w_o \tag{1.18}$$

and that

$$\left(\frac{\Omega}{K}\right)^2 = \frac{1}{\rho \beta} \tag{1.19}$$

Since Ω divided by K gives the wave velocity, c, we have
the wave velocity expressed in terms of the density, ρ,
and the adiabatic compressibility of the material, i.e.,

$$c = \sqrt{1/\rho \beta} \tag{1.20}$$

Equation (1.20) is particularly useful in estimating
the velocity of ultrasound in materials for which no velocity is listed.

For solids, the bulk modulus, B, which is the recip-
procol of the compressibility, β, is more often used. Thus
the velocity of propagation of longitudinal ultrasonic
waves, V(LB), in solids is given by

$$V(LB) = \sqrt{B/\rho} \tag{1.21}$$

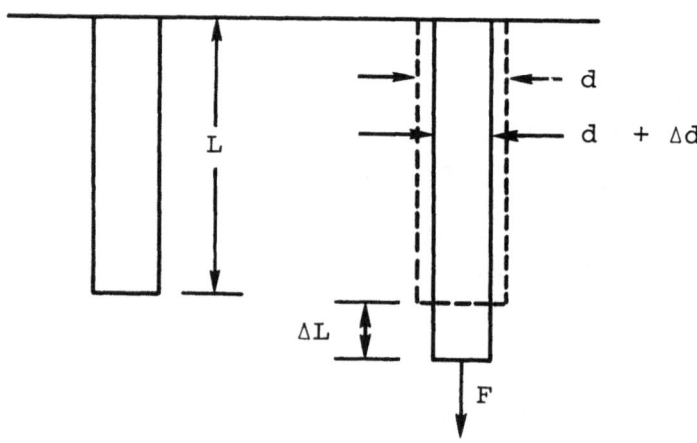

Fig.1.2   The behavior of an elastic rod under tension.

The bulk modulus, B, is related to Young's modulus, E, through Poisson's ratio, σ.

$$\frac{E}{B} = \frac{1-\sigma-2\sigma^2}{1-\sigma} \qquad (1.22)$$

In order to elucidate more clearly the physical significance of Young's modulus and Poisson's ratio, reference is made to Fig. 1.2 illustrating how these quantities are measured. A bar of initial uniform diameter d and initial length L is held at one end by a rigid structure. With the application of a force F to the other end of the bar, the length increases an amount ΔL and the diameter decreases an amount Δd. If the material obeys Hooke's law, the stress, $4F/\pi d^2$, will produce a strain $\Delta L/L$ given by

$$\frac{\Delta L}{L} = \left(\frac{1}{E}\right) \frac{4F}{\pi d^2} \qquad (1.23)$$

i.e., Young's modulus, E, is given by

$$E = \frac{4F}{\pi d^2} \bigg/ \frac{\Delta L}{L} \qquad (1.24)$$

The ratio of the strain, $\Delta L/L$, to the strain, $\Delta d/d$, is known as Poisson's ratio, i.e.,

$$\sigma = (\Delta d/d)/(\Delta L/L) \qquad (1.25)$$

TABLE 1

Calculated and Measured Sonic Velocities
of Two Liquids at $25^\circ C$

Material	$F(CFCF_2O)_5CHFCF_3$ CF_3	Silicone Fluid
Density [kg/m³]	1.79×10^3 *	0.96×10^3
Compressibility [N/m²]	9.58×10^{-10} *	13.1×10^{-10} *
Calculated Velocity [m/s]	764	893
Measured Velocity [m/s]	727	982
*Source	Du Pont Tech. Bull. EL-8B	Dow Corning Bull. 22-069a

Solids will support shear waves as well as compressional waves. In shear waves particle displacements are transverse to the direction of propagation. In Fig. 1.3 the shear stress S may be thought of as being derived from a force couple consisting of two forces F operating in opposing directions and spaced a distance Δz apart. In these terms

$$S = F/\Delta x \Delta y \tag{1.26}$$

If Δz is thought of as equivalent to a length L and the shear strain, γ, as equivalent to $\Delta L/L$ we see that the shear modulus is expressed in a form quite similar to Young's modulus, i.e.,

$$\mu = \frac{F}{\Delta x \Delta y} \bigg/ \frac{\Delta L}{L} \tag{1.27}$$

The velocity of propogation of shear waves, $V(S\mu)$ may be expressed in terms of the shear modulus and, expressed in this form the resulting equation is identical to Eq. 1.21 except that the shear modulus μ replaces the bulk modulus B, i.e.,

$$V(S\mu) = \sqrt{\mu/\rho} \tag{1.28}$$

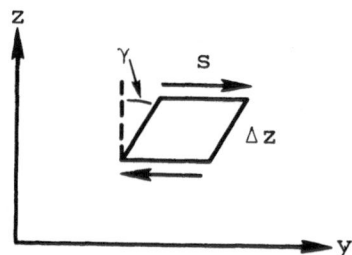

Fig.1.3. An illustration of the parameters used in defining shear strain.

The ratio of the velocity of the longitudinal wave to the velocity of the shear wave in solids may be shown to be[1]

$$\frac{V(LB)}{V(S\mu)} = \sqrt{\frac{2(1-\sigma)}{1-2\sigma}} \tag{1.29}$$

a quantity which is always greater than 1.4. For practically all materials σ lies between 1/4 and 1/2. Using $\sigma = 1/3$ as a typical value for "hard" materials the ratio

$$\left.\frac{V(LB)}{V(S\mu)}\right|_{\sigma=1/3} = 2 \tag{1.30}$$

showing that the shear velocity is always appreciably less than the longitudinal velocity.

1.4 Energy and Intensity

Referring again to Fig. 1.1, we note that the elemental particle of volume $\Delta x \Delta y \Delta z$ is moving with a velocity, U, given by

$$U = \frac{dw}{dt} \tag{1.31}$$

The kinetic energy per unit volume is therefore given by

$$KE = \frac{1}{2} \rho U^2 = \frac{1}{2} \rho \Omega^2 w_o^2 \sin^2(\Omega t - Kz) \tag{1.32}$$

The total energy in the wave is simply the maximum value of the kinetic energy. Thus

$$TE = \frac{1}{2} \rho \Omega^2 w_o^2 \tag{1.33}$$

By intensity we mean the rate of flow of energy through a unit area. Since the wave is propagating with a velocity c, the rate of flow of energy through a unit area, I, is given by

$$I = \frac{1}{2} \rho c \Omega^2 w_o^2 \tag{1.34}$$

Using Eq. (1.9) we find as alternate expressions for Eqs.
(1.33) and (1.34) the following expressions:

$$TE = p_o^2/2pc^2 \qquad (1.35)$$

$$I = p_o^2/2\rho c \qquad (1.36)$$

1.5 Radiation Pressure

An acoustical wave carries with it a certain momentum. When this momentum is changed by reflection or by absorption, a pressure is exerted on the reflector or absorber.
We can calculate this pressure, Π, using the parameters
describing the acoustical wave and the medium in which it
is propagating. In order to derive the required equations
it is necessary to introduce second order effects. Consider, for example, the fact that the density, ρ, of the
material is not constant throughout the volume that the
wave occupies. Rather it varies in accordance with the
instantaneous value of the local pressure, p. To introduce
second order effects we use the expression

$$\rho = \rho_o + \Delta\rho \qquad (1.37)$$

where $\Delta\rho$ represents the increase or decrease in density
when the original volume V is decreased or increased by ΔV.
If the mass of material contained in a volume V is M then

$$\rho = \frac{M}{V} \qquad (1.38)$$

and for sufficiently small changes in volume, ΔV

$$\Delta\rho = -\frac{M}{V^2}\Delta V = -\rho\frac{\Delta V}{V} \qquad (1.39)$$

i.e.,

$$\frac{\Delta V}{V} = \frac{\Delta\rho}{\rho} \qquad (1.40)$$

Equation (1.6) can therefore be modified to

$$\Delta\rho = \rho\beta p \qquad (1.41)$$

Multiplying both sides of Eq. (1.17) by $\sin(\Omega t - Kz)$ it is
easily demonstrated that

$$p = \rho c U \tag{1.42}$$

which, on the basis of Eqs. (1.37) and (1.41) becomes approximately

$$p \approx \rho_o (1 + \beta p) c U \tag{1.43}$$

We now approximate the term βp by

$$\beta p \approx (\rho_o c^2)^{-1} (\rho_o c U) = U/c \tag{1.44}$$

so that

$$p = \rho_o c U + \rho_o U^2 \tag{1.45}$$

We may now identify the radiation pressure as the time averaged value of p. Had we not introduced the second order effect the time averaged value would be zero but the introduction of the U^2- term gives a nonzero value to the integral

$$\Pi = \frac{1}{T} \int_o^T p \, dt \tag{1.46}$$

from which the average value of p is calculated. The period of the wave, T, is given by

$$T = \Omega/2\pi \tag{1.47}$$

and the result of integration is

$$\Pi = \frac{1}{2} \rho_o \Omega^2 w_o^2 \tag{1.48}$$

which, by reason of Eq. (1.34) may also be written

$$\Pi = I/c \tag{1.49}$$

Equation (1.49) yields the pressure on a perfect absorber. At normal incidence, a perfect reflector reverses the direction of momentum so that the rate of change of momentum and therefore the pressure are twice as great for a perfect reflector as for an absorber.

1.6 Velocity Potential

A quantity known as the velocity potential ϕ is often used in dealing with the wave equations of acoustics. The velocity potential is related to particle velocites (not

wave velocities) in the following way. Equation (1.15)
could be generalized to three dimensions by writing

$$-\text{grad } p = \rho \frac{\partial^2 w}{\partial t^2} \tag{1.50}$$

If we define a function ϕ such that

$$p = -\rho \frac{\partial \phi}{\partial t} \tag{1.51}$$

Eq. (1.50) becomes

$$\rho \frac{\partial}{\partial t} \text{grad } \phi = \rho \frac{\partial}{\partial t} \left(\frac{\partial w}{\partial t}\right) \tag{1.52}$$

indicating that

$$\text{grad } \phi = \frac{\partial w}{\partial t} \tag{1.53}$$

A function not dependent upon time could be added to one
side of Eq.(1.53) but, in a liquid, when there is no net
flow, this function is not needed.

In solids, where both longitudinal and shear waves are
present the particle velocity

$$U = \frac{\partial w}{\partial t} \tag{1.54}$$

can be expressed[2] as the sum of a scalar potential ϕ and a
vector potential ψ:

$$U = \text{grad } \phi + \text{curl } \psi \tag{1.55}$$

The potentials ϕ and ψ are the potentials of the longitudinal and shear waves respectively.

1.7 Refraction and Reflection at Liquid-Solid Interfaces

If a plane acoustic wave, unbounded in space or time
is incident upon the interface between a liquid and a solid
medium at an angle $\alpha(IW)$ some of the energy will be reflected
as a shear wave at an angle $\alpha(RS)$, some will be reflected
as a longitudinal wave at an angle $\alpha(RL)$, some will be transmitted as a shear wave at an angle $\alpha(TS)$ and some will be
transmitted as a longitudinal wave at an angle $\alpha(TL)$. The
situation is diagrammed in Fig. 1.4. Fig. 1.4 is drawn for
a general case in which the incident wave, IW, may be in a
liquid or a in solid. No shear wave will be present on the

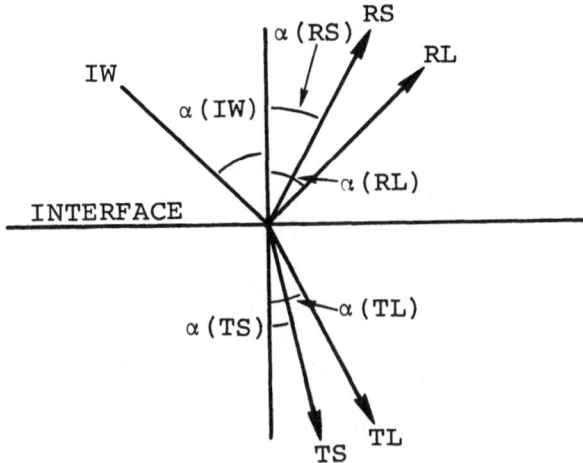

Fig.1.4. Partition of acoustical energy at an interface between two materials.

liquid side. The incident wave may be one of three types. It may be longitudinal or one of two polarizations of the shear wave, one lying parallel and the other lying perpendicular to the plane containing the normal to the interface and the incident ray.

The angles $\alpha(RL)$, $\alpha(TL)$, etc. depend upon the velocities of acoustical waves in the two media. If the indices RL, TL, etc. are used to identify the respective velocities then the well known laws of optics (i.e., Snell's law and the law of reflection) dictate that

$$\frac{V(IW)}{\sin\alpha(IW)} = \frac{V(RS)}{\sin\alpha(RS)} = \frac{V(RL)}{\sin\alpha(RL)} = \frac{V(TS)}{\sin\alpha(TS)} = \frac{V(TL)}{\sin\alpha(TL)}$$

(1.56)

The angles α, are all acute angles measured from the normal to the interface and all the other angles lie on the side of the normal opposite that of $\alpha(IW)$. The medium in which the wave IW is incidence has a density $\rho(I)$. The density of the other medium will be designated $\rho(T)$. Each wave has an acoustic impedance, Z, defined by the equation

$$Z = \frac{\rho V}{\cos\alpha}$$

(1.57)

Appropriate indices will be used to designate each impedance, e.g., the acoustic impedance of the transmitted longitudinal wave is written

$$Z(TL) = \frac{\rho(T)V(TL)}{\cos\alpha(TL)} \tag{1.58}$$

Velocity potential amplitudes, ϕ_O and ψ_O, for each wave are identified simply as RL, TL, etc.

Conservation of energy dictates that

$$\frac{Z(IW)}{Z(RL)}\left(\frac{RL}{IW}\right)^2 + \frac{Z(IW)}{Z(RS)}\left(\frac{RS}{IW}\right)^2 + \frac{\rho(T)}{\rho(I)}\frac{Z(IW)}{Z(TL)}\left(\frac{TL}{IW}\right)^2$$
$$+ \frac{\rho(T)}{\rho(I)}\frac{Z(IW)}{Z(TS)}\left(\frac{IS}{TW}\right)^2 = 1 \tag{1.59}$$

Each term in Eq. (1.59) represents the portion of energy in that wave flowing away from a given area of the interface. Consider the case in which the incident wave is on the liquid side. The incident wave, IW, is designated IL. The reflected shear wave, RS, does not exist. The velocity potential amplitudes may be calculated from the equations

$$\frac{RL}{IL} = \frac{Z(TL)\cos^2 2\alpha(TS) + Z(TS)\sin^2 2\alpha(TS) - Z(IL)}{Z(TL)\cos^2 2\alpha(TS) + Z(TS)\sin^2 2\alpha(TS) + Z(IL)} \tag{1.60}$$

$$\frac{TL}{IL} = \left(\frac{\rho(I)}{\rho(T)}\right)\frac{2Z(TL)\cos 2\alpha(TS)}{Z(TL)\cos^2 2\alpha(TS) + Z(TS)\sin^2 2\alpha(TS) + Z(IL)} \tag{1.61}$$

$$\frac{TS}{IL} = \left(\frac{\rho(I)}{\rho(T)}\right)\frac{2Z(TS)\sin 2\alpha(TS)}{Z(TL)\cos^2 2\alpha(TS) + Z(TS)\sin^2 2\alpha(TS) + Z(IL)} \tag{1.62}$$

$$\frac{RS}{IL} = 0 \tag{1.63}$$

At near normal incidence where $\alpha(IL)$ and hence $\alpha(TS)$ are small, and we see that the transmitted shear wave amplitude TS is much less than the transmitted longitudinal wave amplitude.

Figure 1.5 is a plot of Eq. (1.60) through (1.63) for a water-aluminum interface. At true normal incidence

$$\alpha(IL) = \alpha(TS) = \alpha(TL) = 0 \tag{1.64}$$

and Eqs. (1.60) through (1.63) reduce to

$$\frac{RL}{IL} = \frac{Z(TL) - Z(IL)}{Z(TL) + Z(IL)} \tag{1.65}$$

$$\frac{TL}{IL} = \left(\frac{\rho(I)}{\rho(T)}\right)\frac{2Z(TL)}{Z(TL) + Z(IL)} \tag{1.66}$$

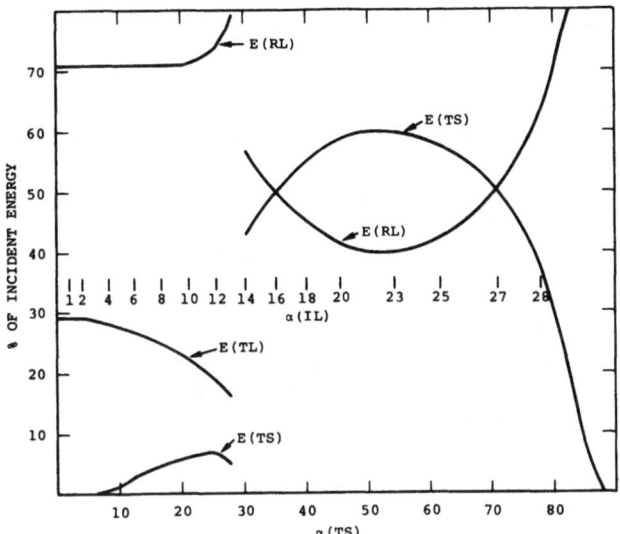

Fig.1.5.   Partition of energy at an interface between water and aluminum.   Incident wave in water.

When a longitudinal wave is incident upon the solid-liquid interface from the solid side the equations giving the relative velocity potential amplitudes are

$$\frac{RL}{IL} = \frac{Z(TL)+Z(RS)\sin^2 2\alpha(RS)-Z(IL)\cos^2 2\alpha(RS)}{Z(TL)+Z(RS)\sin^2 2\alpha(RS)+Z(IL)\cos^2 2\alpha(RS)} \quad (1.67)$$

$$\frac{TL}{IL} = \frac{\cos\alpha(IL)}{\cos\alpha(TL)\cos^2 2\alpha(RS)}\left(1-\frac{RL}{IL}\right)\frac{c}{V(LB)} \quad (1.68)$$

$$\frac{RS}{IL} = \frac{\sin 2\alpha(IL)}{\cos 2\alpha(RS)}\left(1-\frac{RL}{IL}\right)\left(\frac{V(S\mu)}{V(LB)}\right)^2 \quad (1.69)$$

At normal incidence these equations become

$$\frac{RL}{IL} = \frac{Z(TL)-Z(IL)}{Z(TL)+Z(IL)} \quad (1.70)$$

and

$$\frac{TL}{IL} = \frac{2Z(IL)}{Z(IL)+Z(TL)}\left(\frac{c}{V(LB)}\right) \quad (1.71)$$

Finally, there is the case of a shear wave incident from the solid side of a solid-liquid interface. If the particle velocity is parallel to the interface, no longitudinal waves will be generated in either the solid or the liquid upon reflection. This component of the shear wave is totally reflected.

If the particle velocity lies in the plane of incidence, the manner in which the velocity potential amplitude is partitioned may be calculated from

$$\frac{RS}{IS} = - \frac{Z(TL)+Z(RL)\cos^2 2\alpha(IS)-Z(RS)\sin^2 2\alpha(IS)}{Z(TL)+Z(RL)\cos^2 2\alpha(IS)+Z(RS)\sin^2 2\alpha(IS)} \qquad (1.72)$$

$$\frac{TL}{IS} = \frac{\tan\alpha(TL)}{2\sin^2\alpha(IS)}\left(1 + \frac{RS}{IS}\right) \qquad (1.73)$$

$$\frac{RL}{IS} = - \frac{\cos 2\alpha(IS)}{\sin 2\alpha(RL)} \left(1 + \frac{RS}{IS}\right) \left(\frac{V(LB)}{V(S\mu)}\right)^2 \qquad (1.74)$$

At normal incidence

$$\frac{RS}{IS} = -1 \qquad (1.75)$$

and

$$\frac{IL}{IS} = \frac{RL}{IS} = 0 \qquad (1.76)$$

The preceding equations defining the partition of velocity potential amplitudes and energy upon the interaction of an acoustical wave with an interface are basic and needed in the understanding of the interaction but since the description was based upon the assumption of plane waves unbounded in space or time interacting with a single interface they are of limited use for application to most practical physical situations. Practical physical situations usually require an analysis in terms of the linear superposition of many plane waves of differing frequency and direction of propogation. Such an analysis will not be presented here. More detailed and extensive treatments may be found in the references [1,2,9].

Beams of acoustical waves having diameters of a hundred or more wavelengths being formed in wave pockets several hundred wavelengths long can, with a reasonable degree of approximation, be treated as unbounded waves. In practical situations, more than one interface is often encountered within the length of the wave pocket. This gives rise to interference phenomena. One important case is treated in the next section.

1.8   Transmission Through a Thin Plate

The transmission of an acoustical wave at near normal
incidence through a thin plate of solid immersed in water
may be treated in the following way. Only the longitudinal
waves will be considered. The shear wave amplitude is assumed
to be neglibile. If d is the thickness of the plate, if Λ
is the wavelength of the acoustic wave in the solid and if
α(TL) is the angle that the wave in the plate makes with the
normal to the plate as shown in Fig. 1.6 then we define a
phase factor δ such that

$$\delta = \left(\frac{4\pi}{\Lambda}\right) d \cos\alpha \qquad (1.77)$$

Taking the reflection coefficient, r, to be RL/IL at normal
incidence, the intensity of the transmitted wave, I_t, relative to the intensity, I_o, of the incident wave can be calculated from

$$I_o/I_L = 1 + \frac{4r^2}{(1-r^2)^2} \sin\frac{\delta}{2} \qquad (1.78)$$

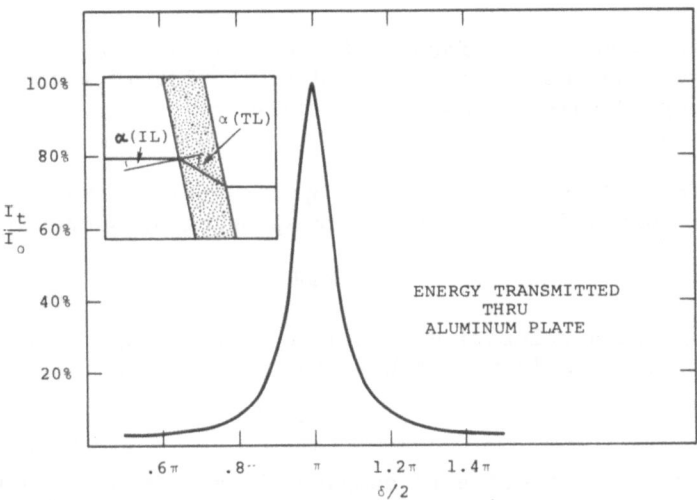

Fig. 1.6.   Transmission of energy through an aluminum plate
in water.

1.9   Absorption

Losses of energy by absorption are handled in the
following manner. If I_d is the intensity of a plane sound
wave at a depth d within the material and if the intensity
at depth d = o is I_o then

$$I_d = I_o \exp(-2\alpha d) \tag{1.79}$$

The units for α are nepers per unit length. When expressed
in decibels per unit length the value is 8.686α. In normal
fluids, in the absence of relaxation phenomena, α varies
as f^2, f being the frequency. In solids α/f is roughly
constant except where scatter and relaxation phenomena exist.
In the soft tissue of biological materials the α/f rule is
also roughly approximated.

1.10   Interaction of an Acoustic Wave with a Free Liquid
        Surface

When a beam of acoustic energy impinges upon a perfect
reflector at normal incidence, a pressure given by (cf.
Eq. (1.49))

$$\Pi = \frac{2I}{c} \tag{1.80}$$

is exerted on the surface. If the beam is several hundred wavelengths in diameter and if the liquid surface extends
well beyond this diameter, the effect of this pressure is to
raise the surface a disance h.

The radiation pressure is opposed chiefly by gravita-
tional forces for which the corresponding pressure is

$$\Pi_g = \rho g h \tag{1.81}$$

g being the acceleration of gravity. Equating the two
pressures and solving for h yields the equation

$$h = \frac{2I}{\rho g c} \tag{1.82}$$

It is instructive to compare the magnitude of surface
displacement as given by Eq. (1.82) with the amplitude of
particle displacement, w_o, which, in accordance with Eq.
(1.34), is given by

$$w_o = \frac{1}{2\pi f} \sqrt{\frac{2I}{\rho c}} \tag{1.83}$$

where the expression $2\pi f$ has been used in place of Ω. For $I = 0.1 \text{watt/cm}^2$, $f = 3\text{MHz}$, $\rho = 1\text{gm/cm}^3$, $c = 1.5 \times 10^5$ cm/sec and $g = 980\text{cm/sec}^2$

$$h = 1.36 \times 10^{-2} \text{cm} \tag{1.84}$$

and

$$w_o = 2.0 \times 10^{-7} \text{cm} \tag{1.85}$$

Thus, for the conditions stated above, the surface displacement caused by radiation pressure is more than four orders of magnitude greater than the particle displacement amplitude.

REFERENCES

1. T. F. Hueter and R. H. Bolt, Sonics, John Wiley and Sons, Inc. New York pp 24-26.

2. L. M. Brekhovskikh, Waves in Layered Media, Academic Press, New York, Chapter 1, Section 4(1960)

3. W. G. Mayer, Energy Partition of Ultrasonic Waves at Flat Boundaries, Ultrasonics, April-June, pp 62-68 (1965)

4. E. Skudrzyk, The Foundation of Acoustics, Springer Verlag, New York (1971)

Chapter 2

HISTORICAL PERSPECTIVES

G. Wade

Electrical Engineering Department

University of California at Santa Barbara

I. INTRODUCTION

To be able to "see" with sound has long been an intriguing concept. Sound waves which are scattered from objects carry much the same image information as do scattered light waves. However, humans are not naturally equipped to efficiently data-process the acoustic image information. A person cannot obtain a good mental image of an object by simply listening to the scattered sound from the object. For example, a man cannot go into a completely dark room and by using his vocal chords only, obtain much information, if any at all, that will give him a mental image of the objects in the room. He may shout and listen for echoes, but to almost no avail. On the other hand, he can do precisely the equivalent thing with light. If he takes a flashlight into the room, he can quickly turn it on and off, creating flashes of light. By seeing the pulses of scattered light from the various objects he can readily obtain a mental image of the size, color and configuration of the various parts of the room and of the objects placed within it. The human ear will not data-process rapidly and effectively enough to provide a mental image of scattering centers for sound, but the human eye will do exactly that as far as light is concerned.

Although nature has not equipped man to "see" with sound, certain animals can do this very well. The bat, for example, is an expert at it. Bats can fly about in a completely darkened room and avoid hitting each other, the

walls, the floor or ceiling or any of the ordinary objects
in the room. Obviously, the evolutionary development
of the bat has provided that animal with a high level of
sophistication in its acoustic imaging capability. The
bat's talent in using sound is comparable to that of other
animals in using light. Bats have a specialized larynx
that produces high-energy ultrasonic squeaks. A relatively
large portion of the bat's brain is devoted to processing
ultrasonic echo information, just as a considerable part
of the brain of other animals is used for vision.

Only recently have these facts been fully appreciated.
In 1793, the Italian scientist Spallanzani wrote that
he was inclined to believe that bats have "a new organ or
sense" that humans do not have and are not even aware of [1].
It was not until 1920 that the principle of operation of the
"new sense" was found to be somewhat similar to that used
by the sonar systems developed a few years before, during
World War I.

By now it is well known that several cetaceans, such
as the dolphin, also have excellent capability for navigat-
ing and locating food by means of sonic energy. The dolphin,
interestingly enough, has very good eyes, but it nevertheless
depends primarily on its ability to process acoustic echoes
for locating food. These animals in captivity have almost
no objection to a blindfold for their eyes, but they strongly
resist having their sense of hearing impaired in any way
whatsoever.

It is a curious fact that historically, of all the
known forms of radiation, acoustic waves have been among the
last to be exploited by humans for producing images. Many
kinds of systems and devices for imaging have been invented,
such as those using visible light, X rays, infrared, micro-
waves, and even electron beams. However, only within the
last 50 years has man discovered that he can overcome
natural limitations in "seeing" with sound by exercising
ingenuity. Since then, a multiplicity of methods have been
proposed for providing such a capability.

II.   EARLY PIONEERS

2.1 Langevin

One of the first solid accomplishments in using
ultrasound for "imaging" was made by Paul Langevin during
World War I. Langevin, a major French physicist, had
widely ranging interests and had previously studied X rays,
relativity, para-magnetism, dia-magnetism, ionic transport,
and thermodynamics, among other scientific topics. Although
he concerned himself mainly with philosophical questions,
he did not neglect the technological implications of his
work. One of the practical problems faced by the French
during World War I was how to detect submerged enemy sub-
marines, and Langevin was asked to find a way to do so.
Both Lord Rayleigh and O. P. Richardson had previously
thought of employing ultrasonic waves. An engineer,
M. C. Chilowski, developed an ultrasonic device for the
French Navy, but its acoustic intensity was much too weak
to be practical. Langevin looked into the question of how
to increase the acoustic power output and in less than
three years succeeded in providing high ultrasonic intensity
by means of a piezoelectric transducer operating at
resonance [2]. Langevin continued to do important work
in acoustics and ultrasonics even after the war, and as
late as 1940 he was invited by the French Navy to direct
a research program on ultrasonic depth-finders.

2.2 Sokolov

Acoustic "imaging" by sonar was just one of the many
scientific and technical interests of Langevin. In fact,
Langevin's work in this area is by no means the principal
reason for his being remembered today. However, in the
case of another great European scientist, S. J. Sokolov of
the University of Leningrad, ultrasonic imaging was indeed
the principal pursuit and reason for renown. Sokolov's
productive work extended over three decades, beginning in
the 1920's. He was one of the first persons to recognize
and systematically explore the usefulness of ultra-sound
for "imaging" internal structure in optically opaque
objects. He devised several techniques for producing
optical patterns corresponding to metal objects that were
irradiated with sonic beams in the megahertz region. Many
of his schemes were proposed in order to detect inhomo-

geneities, such as flaws and voids, within the objects.

In one of his systems, the inhomogeneities were made
"visible" by reflecting collimated light from a liquid
surface in a fashion similar to that of liquid-surface
holography. In another system, light was diffracted after
being passed through a glass container filled with turpen-
tine in a manner reminiscent of Bragg-diffraction imaging.
The recent work on Bragg-diffraction imaging and liquid-
surface holographic imaging goes back only to the middle
of the last decade. But the ideas of Sokolov were quite
similar in nature to these modern ideas and were enunciated
more than 30 years earlier.

The liquid-surface system he constructed for detecting
flaws in metal test pieces is illustrated in Fig. 1. The
scheme was first proposed in 1929. Later Sokolov patented
the system and published a paper concerning its operation [3].
This system is remarkably like the modern liquid-surface
systems described in detail in Chapter 6 and commented
upon briefly later in the present chapter. However, two
major differences exist. Modern systems have a second sound
source referred to as a reference beam. They also use lasers
to read out the image information. Lasers, of course, were
not available to Sokolov, and his light source was not as
coherent as he might have wished. However, the absence of
a second source does not change the character of the operation of the system in any fundamental way. Even without
the second sound source, the formation of static ripples
could be expected to take place on the oil surface of the
system in such a way as to form a hologram. Flaws of
sufficiently small dimensions would obviously permit the
passage of a substantial portion of undisturbed sound
through the metal piece. This sound can be regarded as
constituting a reference beam. The object beam is then
that part of the incoming sound scattered from the flaws.
Thus we have the acoustic equivalent of the "Gabor-type"
hologram of optical holography.

Conceivably, therefore, Sokolov can be regarded as
having produced the first hologram, and doing so a number
of years before Gabor's invention of holography. It is
therefore possible to argue that, from a chronological
standpoint, acoustic holography preceded optical holography.
Of course, Sokolov had no way of understanding holographic

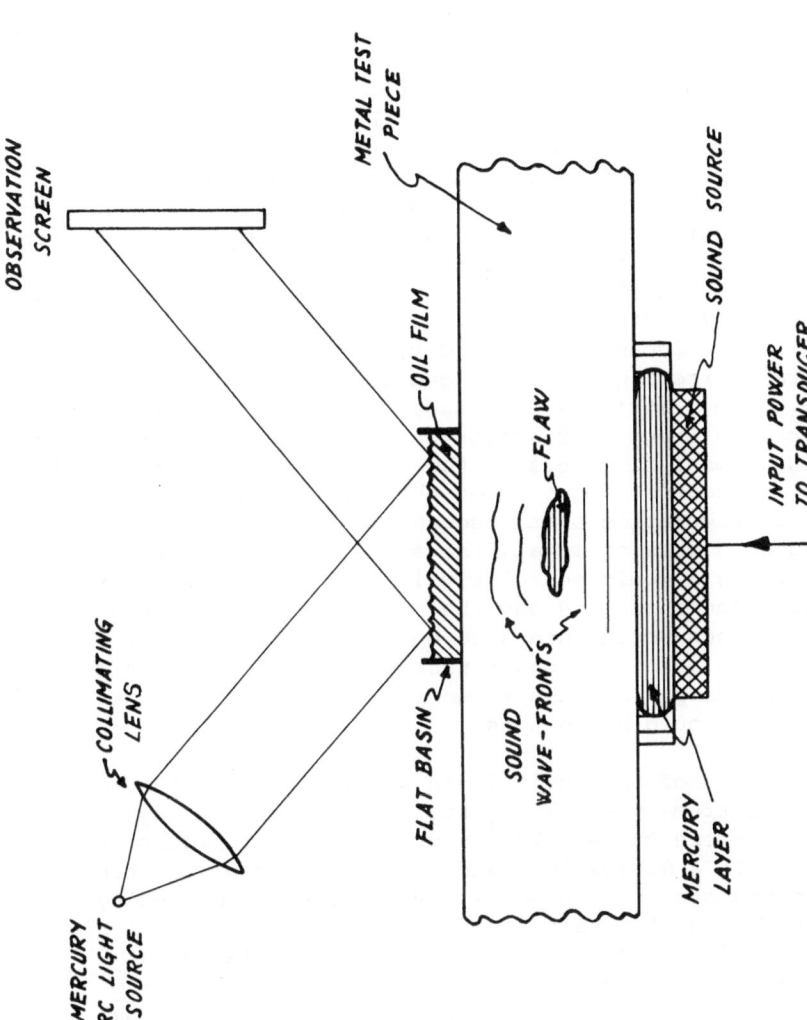

Fig. 1 Sokolov's liquid-surface system for detecting flaws in metal test pieces.

principles as we understand them currently. His images
were not of high quality. Not only did he operate
without the use of a laser, but also he provided no
spatial filtering to eliminate unwanted beam components
in the reconstruction process.

In another of Sokolov's early attempts to detect flaws
with ultrasound, he used a bulk system strikingly similar
in overall appearance to that of the present Bragg-diffraction systems [4]. A diagram of Sokolov's system is shown
in Fig. 2. Compare this system with the Bragg-diffraction
systems described in Chapter 9 and illustrated in Fig. 2
of that Chapter. Again, Sokolov was interested in detecting
inhomogeneities, such as cracks and casting errors, in
manufactured metallic parts, rather than in forming images
per se. His system utilized the arrangements of Debye and
Sears and of Lucas and Biquard to display the diffraction
spectra of the light as produced by interaction with the
sound. Thus, strictly speaking, Debye-Sears diffraction
rather than Bragg diffraction was involved. The intensity
and the number of orders of the diffraction spectra displayed
would, of course, depend upon the strength and the wavefront configuration of the sound in the metal. For a
homogeneous mass of metal, the sound power reaching the
acoustic cell would be relatively high and the wave fronts
planar. The intensity and the number of orders appearing
on the screen would then also be high. However, for a non-
homogeneous mass, the sound passing through would be scattered and damped and the intensity and number of orders on
the screen would be reduced.

Another of Sokolov's ideas was initially intended for
imaging extremely small objects. He proposed using sound
at 3 GHz where the wavelength of the sound in water is very
short (half a micrometer) and capable of resolving truly
minute objects. The name he gave to the corresponding
device was the "ultrasonic microscope" [5]. Technological
impracticalities prevented Sokolov from operating at
sufficiently high frequencies (such as 3 GHz) for the
purposes of microscopy. However, the principle he put
forth (that of reading out localized electronic charge
developed on a piezoelectric crystal in response to an
acoustic input) has since become embodied in a well-known
device called the Sokolov tube which is used for low-
frequency acoustic imaging [6].

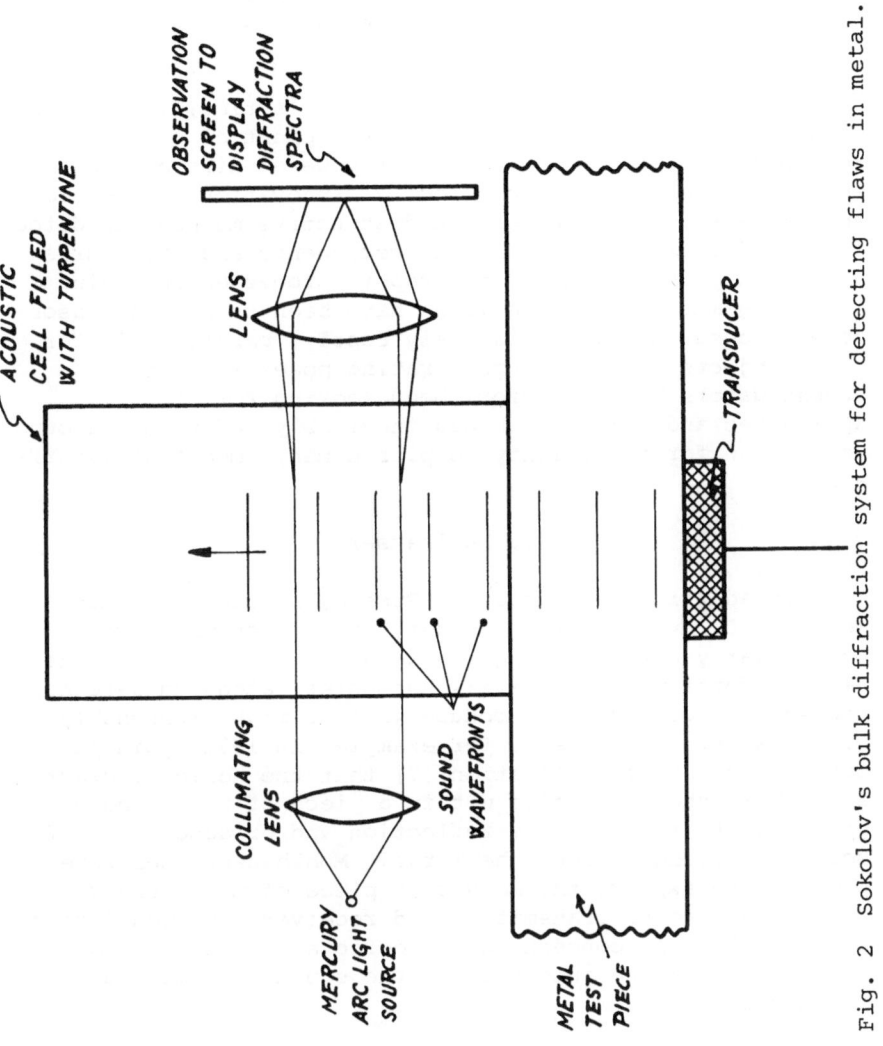

Fig. 2 Sokolov's bulk diffraction system for detecting flaws in metal.

In terms of the quantity of his ideas, the quality of
his accomplishments and the extended period of time over
which his explorations took place, Sokolov can be called
the principal pioneer in acoustic imaging. As we have seen,
much of his motivation stemmed from a desire to produce
systems that would carry out what we now call non-destructive
testing (NDT).

As important as NDT was in Sokolov's time, it is even
more important today. With the development of nuclear
reactors, deep-ocean vehicles, jet-powered aircraft, manned
spacecraft, and other technological innovations of the last
thirty years, the need for non-destructive methods to detect
flaws and inclusions in various components and structures
has greatly increased. Radiographic processes involving
X rays, gamma rays and neutrons have been used with reasonable success, but are relatively costly, particularly for
thick objects where high-penetrating power is needed.
Investigators have therefore been looking for other
approaches and this search has inevitably led to ultrasound,
well-known for its ability to pierce many important materials.

2.3 Muhlhauser

In addition to Sokolov in Russia, a trio of German
workers, O. Muhlhauser, A. Trost, and R. Pohlman were
particularly active during the decade of the 1930's. Their
early efforts were relatively unsophisticated and were not
necessarily designed to produce what we might reasonably
call images. Muhlhauser, for example, in 1931, pointed
out in a patent specification [7] that the hollows, cracks,
and other inhomogeneities within a piece of metal would
give rise to a noticeable reflection and/or absorption of
sound traveling through the metal. Muhlhauser therefore
proposed to explore the mass of a piece of metal under
test with a sound transmitter and receiver. He would then
draw conclusions concerning the defects in the interior of
the metal from the intensity of the sound transmitted
through it.

Obviously Muhlhauser's approach was extremely crude.
It could be relied upon to give an indication, although
certainly not an image, of interior inhomogeneities. However, its operation was instantaneous, and Muhlhauser's work

was the forerunner of more sophisticated efforts in real-
time acoustic imagery.

2.4 Pohlman

Reimar Pohlman, a fellow German, was soon to propose
more effective techniques for instantaneous ultrasonic
imaging. One of Pohlman's ideas, embodied in what is
presently called the Pohlman Cell, was first described in
1937 [8]. The cell consists of a sandwich containing a
suspension of fine metallic flakes in a suitable liquid.
As shown in Fig. 3, one side of the sandwich is a glass
plate. The other side is a thin membrane which is
acoustically transparent. Because of thermal motion, the
metallic flakes will normally be oriented randomly in the
medium. If the suspension is illuminated by light projected
through the glass plate, then the light reflected from the
particles as seen by an observer will have a diffusely
scattered matte appearance. However, when sound, passing
though the stretched thin membrane, is incident on the
suspension, the metallic flakes will act as miniature
Rayleigh discs and tend to align themselves locally with
their surfaces parallel to each other. Each flake is
inclined to set itself in such a way as to present maximum
area to the sound field. Thus the plane of each flake in
regions of high acoustic intensity tends to become perpendicular to the direction of the acoustic propagation. Then if
the suspension is irradiated by light, the reflected light
delineates the pattern of acoustic intensity impinging
upon the cell through the membrane on the other side.
The regions of high acoustic intensity with their aligned
metal flakes will present a more reflective surface to
the incident light than the other regions with their
randomly oriented flakes. Under these conditions the eye
of an observer will see bright areas, corresponding to the
regions of non-zero acoustic intensity, superimposed on
the gray matte appearing background, corresponding to the
regions of zero acoustic intensity. The degree of brighten-
ing in the various bright areas will depend upon the level
of the acoustic intensity.

The Pohlman cell has relatively high intrinsic resolution
because the size of the metallic particles used in the suspension and the thickness of the membrane through which the
sound passes can be very small compared with a wavelength.

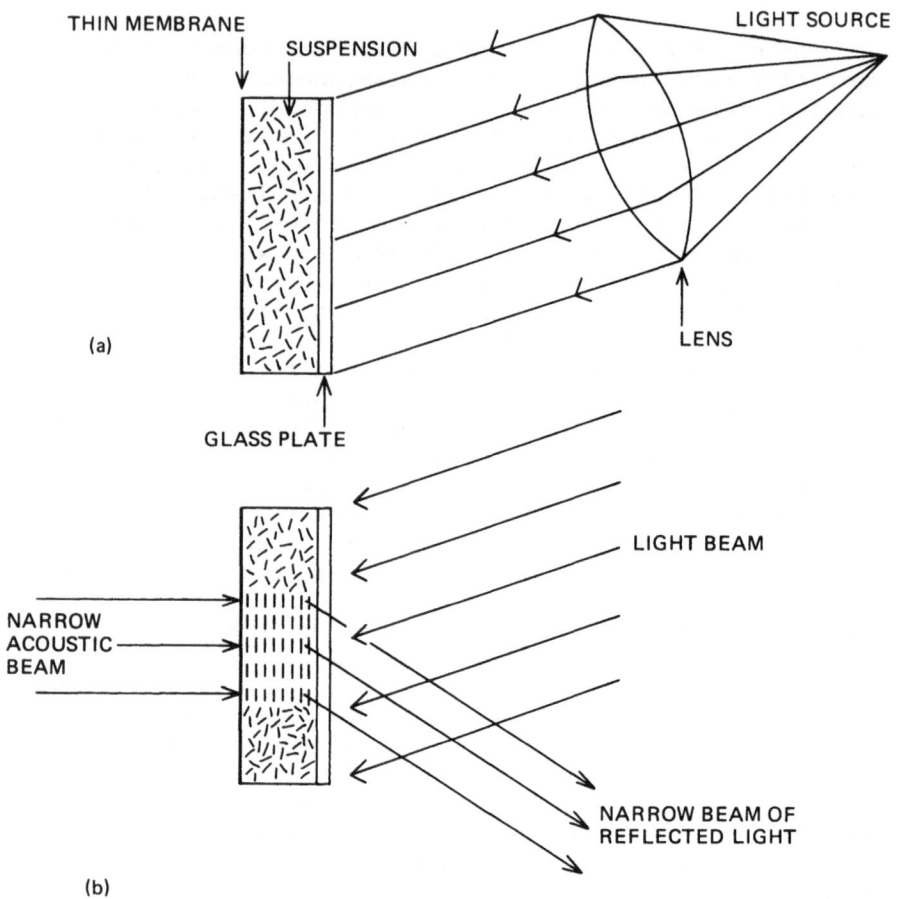

Fig. 3   Diagrams illustrating how the Pohlman cell works.
         (a) cell with no insonification.
         (b) cell with localized insonification.

The main disadvantage of this imaging technique is that the
response time for the particles to orient themselves when
insonification is present, or to become randomly oriented
in the absence of insonification, is too long for authentic
real-time operation. Nevertheless, the idea is intriguing
and has attracted much attention over the years since it
was first proposed. In fact, the Pohlman cell is still
being studied at present in various research laboratories
for a variety of specialized applications.

2.5 Firestone

A great advance in non-destructive testing by ultrasound was made very shortly after Pohlman's work by F. A.
Firestone of the University of Michigan, who invented an
instrument called the Ultrasonic Reflectoscope. The
antecedent for Firestone's invention was Langevin's sonar
system dating back to World War I. Firestone's Reflectoscope became the forerunner of all modern pulse-echo test
systems. It operated by generating a succession of ultrasonic pulses of short duration and detecting their echoes
from the sub-surface discontinuities within the solids. The
Reflectoscope made it possible to examine an object from one
surface only, to determine within limits the size of any
internal inhomogeneities and to measure the depth of the
inhomogeneities below the surface.

Ever since Firestone's day the pulse-echo approach has
had substantial success in non-destructive testing. As
with radar, the systems based on this approach provide a
map of the regions under investigation, producing, by means
of the so-called B-scan, cross-sectional "images" of the
internal structure of the object. These "images" are some-
times referred to as "artificial images." They are not of
the true-perspective type, that is, they are not orthographic in character, but rather are like the sector scans
obtained with radar. In addition to their being used in
NDT, pulse-echo systems are also being employed in medical
science where the displays they produce are frequently
called tomograms. Hence these systems are proving to be
useful, not only in non-destructive testing, but also in
medical diagnosis.

The pulse-echo approach works particularly well for
objects of small area. Large-area coverage by such a

system requires complex mechanical scanning, or the use of
numerous transducers in an array. Operation in real time
is also a problem. Therefore, other methods of acoustic
imaging, especially where real-time performance is desired,
offer attractive alternatives.

III. HOLOGRAPHIC APPROACHES

Since the days of Sokolov, Pohlman, and Firestone a
multiplicity of other methods have been proposed both for
testing materials non-destructively and for medical diagnosis. The invention of holography and the advent of the
laser have, in recent years, led to renewed interest in
ultrasonic imaging, and to the improvisation of new acousto-
optic techniques in this regard. Dennis Gabor, inventor
of holography and recent Nobel Laureate in Physics, said
some time ago [9], "I see an important application of vision
by ultrasound in medical diagnostics, where it could not
only replace X rays, but score above them in making visible
fetuses, clogged veins and arteries, and incipient tumors."
A number of other investigators have thought along similar
lines and throughout the last decade there has been considerable research activity in this general area, much of it
reminiscent of the principles and ideas enunciated by the
early pioneers in acoustic imaging previously mentioned.
However, the impetus for this recent work was not inspired
so much by what the pioneers did as by the advances in
optical holography. In fact, many of the new systems
employ laser beams and are either holographic in nature
or can be thought of in terms of holography. By now, a
wide variety of system concepts exist, and several of them
have demonstrated astonishing capability.

3.1 Sonographic Plate

The first person to produce and report on an ultrasonic hologram was Pal Greguss, in 1965 [10]. Greguss'
method for recording a hologram was fundamentally different
from any of the approaches presently being used and illustrates the ingenuity and variety of thought that has been
employed in this field during the last decade. To record
the hologram, Greguss used a "sonographic plate," whose
exposure depended upon a "sonochemical" reaction rather
than on a photochemical one. If such a plate is developed
in the presence of a sound field, the pattern of standing-

wave components in the sound will be converted into a black
and white image suitable for use in making a hologram. This
approach has serious drawbacks because the sound used for
the recording process must be very intense.

3.2 Point-by-Point Scanning Systems

The type of holographic system which, from the publication dates in the research literature, has been investigated
over the longest sustained period of time, is the kind that
uses a scanning, detecting transducer moving through a
raster pattern in the holographic plane. The procedure
involved was first described in an elementary form by
Thurstone, in 1966 [11]. A typical system is illustrated
in Fig. 4. Normally, water is used as the medium for
acoustic propagation. In the embodiment shown, a sound
source is placed on one side of the object, and a scanning
hydrophone on the other. The effective diameter of the
hydrophone should be less than, or at most equal to, the
fringe spacing to be recorded. This kind of system will
be described in detail in Chapter 5.

3.3 Liquid-Surface Systems

A second holographic approach has also been the subject
of much sustained research over the past few years. The
principles of operation were enunciated by Mueller and
Sheridon in their classic paper of 1966 [12]. The system
uses a liquid-air interface as the recording medium and
thus is similar in this regard to Sokolov's system of Fig. 1.
Fig. 5 shows the arrangement. The scattered wave fronts of
the object beam move upward toward the surface of the water
in the tank, as do the relatively well-ordered wave fronts
of a reference beam. In this system, two different transducers are used to provide the object beam and the reference
beam, whereas only one source is used in Sokolov's system.
The wave fronts from the reference and object beams undergo
reflection at the liquid-air interface above the membrane
shown in the picture. This reflection produces radiation
pressure on the liquid surface. The pressure causes the
surface to deform until gravity and surface tension achieve
a new balance. The deformation shows up as a stationary
ripple pattern corresponding to the interference between
the two beams. The ripples can be used directly to produce
spatially modulated, first-order diffraction side bands on

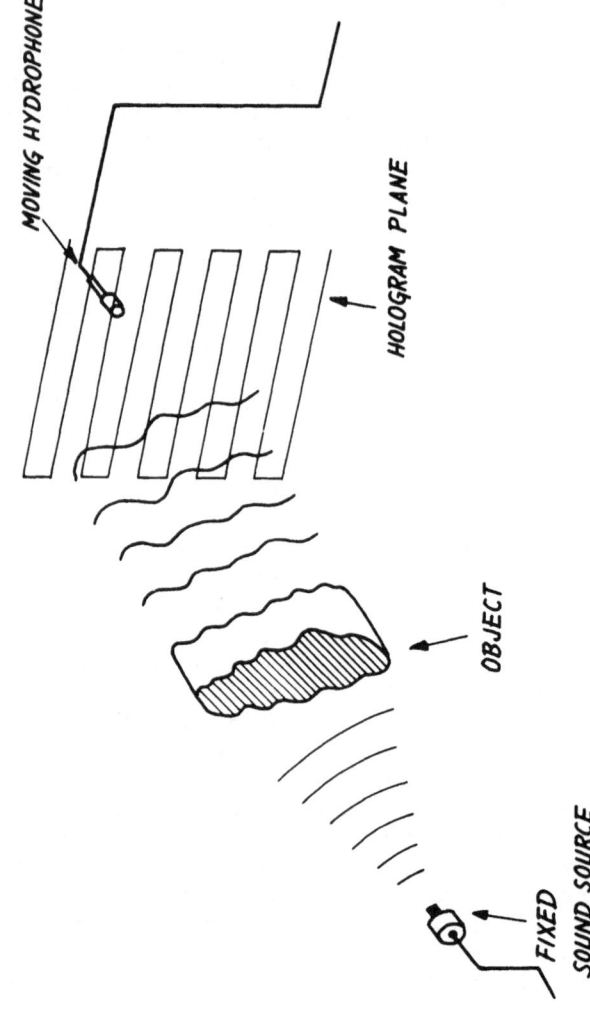

Fig. 4 An acoustic holographic system using a scanning,
detecting transducer. Note that no acoustic
reference beam is needed since the reference can
be simulated electronically.

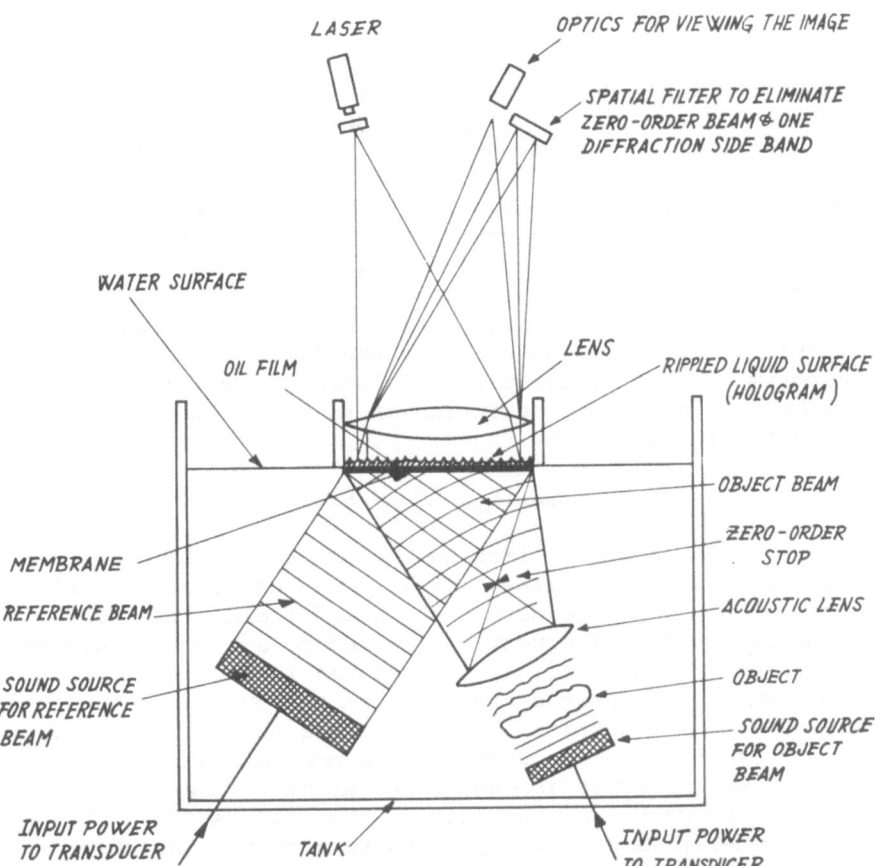

Fig. 5 Liquid-surface holographic system. Static ripples
        on the liquid-air interface constitute a record of
        the interference between the object and reference
        beams and permit the rippled liquid surface to
        operate as a hologram in real time.

a beam of laser light. Thus the ripple pattern on the
liquid surface becomes the hologram of the object. The
zero-order light and one of the diffraction side bands are
eliminated by an optical spatial filter before the image is
further processed for viewing.

This type of holography is frequently referred to as
"liquid-surface" holography, or "static-ripple" holography.
Image reconstruction takes place instantaneously. The best
images are achieved when an acoustic lens is placed between
the object and the surface, so that an acoustic image is
formed directly on the hologram plane (as illustrated in
Fig. 5). Under these circumstances the reconstructed visual
image can be of extremely high quality. In fact, images
produced this way have been among the best obtained to date
by any of the acoustic imaging systems. A hologram made
in this fashion is equivalent to the "focused-image"
hologram, or the "image-plane" hologram of optical holography. A detailed description of this type of system is
presented in Chapter 6.

IV. SOLID-SURFACE SYSTEMS

A second acoustic imaging system that employs laser-
beam readout and operates in real time, utilizes a solid
surface from which to detect the sound field. The technique involved is frequently referred to as "solid-surface"
holography. The system is illustrated in Fig. 6. The
sound waves are first scattered from the object to be
viewed into the water bath in which the object is placed.
The waves then impinge upon a solid membrane with a shiny
surface which serves as one of the sides of the tank containing the bath. As they strike the solid surface, the
waves generate a moving ripple pattern on the interface
between the membrane and the water. Even though no
reference beam is used in the system, the surface records
what might be called a "dynamic hologram." The image is
read out by using a scanning, focused laser beam which
produces a spot on the surface. The spot scans through a
raster pattern on the "holographic" plane, and is reflected
from the shiny surface. In the process, the beam becomes
angularly modulated by the dynamic ripple on the surface.
The image information is converted into weak electrical
signals by passing the reflected laser beam through a
knife-edge-and-photodiode combination. The signals are

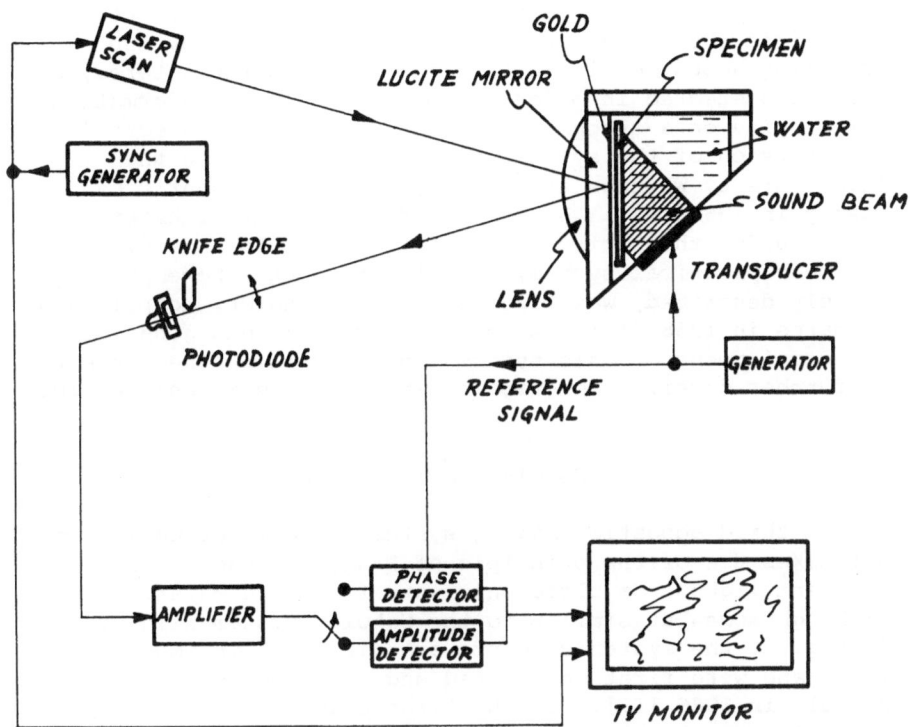

Fig. 6   Solid-surface holographic system.   Image information
         contained in the dynamic ripples on the surface of
         the lucite mirror is read out by a scanning laser
         beam.

amplified and decoded, resulting in an acoustic micrograph
displayed on a synchronously-scanned television screen.
This system has recently been packaged by Sonoscan, and
is presently commercially available for acoustic microscopy [13].

An earlier, somewhat similar version of this approach
operated in a slightly different fashion. The reflected
light from the shiny surface is Doppler-shifted by the
traveling acoustic ripple, and may be optically heterodyned
and photodetected in order to obtain the image information
for display on a synchronously-scanned television screen.
This version of the system was first described in the
literature by Whitman et al. [14], and independently by
Massey in 1968 [15]. However, efficient optical hetero-
dyning under these circumstances is difficult to achieve.
The more practical system, with the readout scheme previously described, was first proposed by Adler, Korpel, and
Desmares in 1968 [16]. Much effective work has gone into
the development of this system, and it will be the subject
of further detailed description and analysis in Chapter 10.

V.   BRAGG-DIFFRACTION SYSTEMS

A third acoustic imaging system using laser-beam readout makes use of the principle of Bragg-diffraction of
coherent light from ultrasonic wave fronts in water.
Although somewhat similar to the Sokolov system illustrated
in Fig. 2, the system concepts are different in important
ways, and were first enunciated and experimented with by
Korpel* in 1966 [17]. In the first place, a wedge-shaped
beam of laser light is caused to traverse the acoustic
cell rather than a collimated beam from a mercury arc as
in the Sokolov system. This is illustrated in Fig. 2 of
Chapter 9. In the second place, Bragg diffraction is
used rather than that of Debye and Sears. In the third

---

*   The ideas involved were independèntly thought of by
     Hance, Parks, and Tsai [18] at Lockheed Research
     Laboratories, and by the present author [19] at UCSB.
     Research efforts similar to Korpel's at Zenith were
     mounted at about the same time by these workers.

place, a masking stop is inserted in the system (see the above-mentioned figure) to prevent unwanted light from interfering with the image.

The techniques employed in Bragg-diffraction systems are sufficiently different from those used in the more clearly holographic systems that many workers in the field of acoustic imaging do not regard Bragg-diffraction systems as being holographic at all. For example, with a Bragg-diffraction system there is no need for either an acoustic reference beam or its equivalent in order to retain phase information in the image. This information is conserved automatically in the Bragg-diffracted light that leaves the sound cell. This and other aspects of the system and its relation to holography are described in detail in Chapter 9.

VI. SYSTEMS USING PIEZOELECTRIC READOUT

Perhaps the most remarkable images of all have been obtained by ultrasonic cameras and microscopes utilizing piezoelectric readout rather than laser-beam readout. Some of these piezoelectric systems use arrays of sonic receiving and/or transmitting elements. Others employ acoustic lenses, either with or without the arrays. Systems in this category can be put together in a variety of ways, a number of which are described in detail in Chapters 7, 8, and 11.

By and large, these systems are not basically holographic in nature, although they can readily be modified to function in a holographic mode if this is wanted. However, in their typical operation the necessary focusing is accomplished either through the use of acoustic lenses or arrays of acoustic elements, and the image information is read out by means of piezoelectric transducers. In terms of sensitivity and freedom from certain spurious image elements (speckle and ringing), this latter category of systems is inherently superior to the category of systems using laser-beam readout. These points will be discussed in more detail in Chapter 4. A number of exceptional images have been obtained from systems in this category, notably from an ultrasonic camera operating at low frequencies, developed by Philip Green and colleagues at Stanford Research Institute [20], and from an acoustic microscope operating at high frequencies developed by Professor Quate at Stanford University [21].

VII. CONCLUSION

From this brief chronological review, we can see that acoustic imaging has previously attracted and is presently attracting the attention of a number of workers. The attempt to use sound for imaging possesses a history that goes back a number of decades. However, over the past ten years there has been a spurt of new research activity in this field, stemming primarily from the advent of the laser and the strides in optical holography and concentrating particularly on orthographic imagers as opposed to sector-scan imagers.

The work in real-time, true-perspective imaging with sound has matured to the extent that several systems are presently being examined in clinical or industrial settings to assess their value in various practical ways. The period of rapid experimental achievement involving new, original ideas and solid theoretical development seems to have already peaked out, and is perhaps now on the decline to some extent, but a new period of developmental activity and of careful and detailed examination of practical applications and potentialities is in the offing.

There are still a large number of interesting and important problems to be solved. Solutions to these problems must be provided and improvements must be brought about before the full potential of acoustic imaging will be realized in practical situations. There are many unanswered questions and unexplored techniques to be checked out. Obviously much work remains to be done. Nevertheless, this work is being done, and it now appears more and more likely that someday we will be able to look back at the present activity and say that it was well worthwhile. From what has already been accomplished, we can reasonably expect that the new systems will find employment not only in medical diagnosis, acoustic microscopy and non-destructive testing, the areas emphasized in this chapter, but also in such applications as oceanic search and seismic sensing, areas which have not previously been mentioned here.

VIII.   REFERENCES

[1]   L. Spallanzani, "Litera Prima," in Opere de Lazzaro
      Spallanzani, Vol. 5, Dalla Societa Tipogr. De'Classici
      Italiani, Milano, 1826, p. 209.

[2]   P. Langevin and M. C. Chilowski, "Procédés et
      appareils pour la production de signaux sous-marins
      diregés et pour la localisation a distance d'obstacles
      sous-marins," French Patent No. 502913, May 29, 1916.

[3]   S. J. Sokolov, "Ultrasonic oscillations and their
      applications," Tech. Phys. U.S.S.R., Vol. 2,
      p. 522, 1935.

[4]   S. J. Sokolov, "Über die praktische ausnutzung de
      beugnung des lichtes an ultraschällwellen," Phys. Z.,
      Vol. 36, p. 142, 1935.

[5]   S. J. Sokolov, "Ultrasonic microscope," Akedemia Nauk
      SSSR; Doklady (Tekhnicheskaya Fizika), Vol. 64,
      pp. 333-335, 1949.

[6]   J. E. Jacobs and D. A. Peterson, "Advances in the
      Sokoloff tube," in Acoustical Holography, Vol. 5,
      P. S. Green, Ed. New York: Plenum, 1974, pp. 633-645.

[7]   O. Muhlhauser, "Verfahren zur Zustandsbestimmung von
      Werkstoffen, besonders zur Ermittlung von Fehlern
      darin," German Patent No. 569598, 1931.

[8]   R. Pohlman, "Über die richtende Wirkung des Schallfeldes
      auf Suspensionen nicht kugelförmiger Teilchen,"
      Zeitschrift für Physik, Vol. 107, pp. 497-507, 1937.

[9]   Editorial Staff, "An interview with the father of
      holography," Optical Spectra, Vol. 4, No. 9,
      pp. 32-33, October 1970.

[10] P. Greguss, "Ultrasonic holograms," Res. Film, Vol. 5:4
      pp. 330-337, 1965.

[11] F. L. Thurstone, "Ultrasound holography and visual
      reconstruction," Proc. Symp. Biomed. Eng., Vol. 1,
      pp. 12-15, 1966.

[12]   R. K. Mueller and N. K. Sheridon, "Sound holograms
       and optical reconstruction," Appl. Phys. Lett.,
       Vol. 9, pp. 328-329, November 1966.

[13]   L. W. Kessler, P. R. Palermo and A. Korpel, "Practical
       high-resolution acoustic microscopy," in Acoustical
       Holography, Vol. 4, G. Wade, Ed. New York: Plenum,
       1972, pp. 51-71.

[14]   R. L. Whitman, L. J. Laub, and W. J. Bates,
       "Acoustic surface displacement measurements on a
       wedge-shaped transducer using an optical probe
       technique," IEEE Trans. on Sonics and Ultrasonics,
       Vol. SU-15, pp. 186-189, July 1968.

[15]   G. A. Massey, "An optical heterodyne ultrasonic
       image converter," Proc. IEEE, Vol. 56, pp. 2157-2161,
       Dec. 1968.

[16]   R. Adler, A. Korpel and P. Desmares, "An instrument
       for making surface waves visible," IEEE Trans. on
       Sonics and Ultrasonics, Vol. SU-15, pp. 157-161,
       July 1968.

[17]   A. Korpel, "Visualization of the cross-section of a
       sound beam by Bragg-diffraction of light," Appl. Phys.
       Lett., Vol. 9, pp. 425-427, Dec. 1966.

[18]   H. V. Hance, J. K. Parks and C. S. Tsai, "Optical
       imaging of a complex ultrasonic field by diffraction
       of a laser beam," J. Appl. Phys., Vol. 38, No. 4,
       pp. 1981-1983, March 1967.

[19]   G. Wade, J. Landry, and A. A. deSouza, "Acoustic
       transparencies for optical imaging and ultrasonic
       diffraction," paper at the First International
       Symposium on Acoustical Holography, December 1967.

[20]   P. S. Green, L. F. Schaefer, E. D. Jones and
       J. R. Suarez, "A new, high-performance ultrasonic
       camera," in Acoustical Holography, Vol. 5,
       P. S. Green, Ed. New York: Plenum, 1974, pp. 493-503.

[21]   R. A. Lemons and C. F. Quate, "Acoustic microscope-
       scanning version," Appl. Phys. Lett., Vol. 24, No. 4,
       pp. 163-165, Feb. 1974.

Chapter 3

INTRODUCTION TO ACOUSTIC IMAGING SYSTEMS

Lawrence W. Kessler

Sonoscan, Inc.
752 Foster Avenue
Bensenville, Illinois 60106

Abstract

The purpose of this and the next chapter is to describe some of the basic concepts and systems for acoustic imaging. Consideration is given to modes of operation of imaging systems and to the variety of methodologies employed for implementing systems. Separate chapters of this book include more detailed accounts of some of the more important (or popular) techniques receiving attention today. Therefore, in this chapter topics will be presented in a rather introductory fashion in order to assemble the concepts for ultrasonics imaging system analysis and synthesis.

Basic Imaging Systems

Figure 1 illustrates the two basic configurations for imaging an object, through transmission and reflection. A source of illumination, or insonification, is aimed in the general direction of an object. Collimated energy from the source is depicted for the purpose of illustration. This happens to be a common technique in ultrasonic visualization, however, it is not necessarily ideal or common for other methods of visualization. Referring to Figure 1, the angular distribution of the reflected and the transmitted energy may substantially differ from that of the incident. The total angular spread defined as $2\theta_r$ and $2\theta_t$, respectively, depends upon the size, shape, and orientation of structural details on and within the object. The strength of the scattered energy depends upon one additional factor,

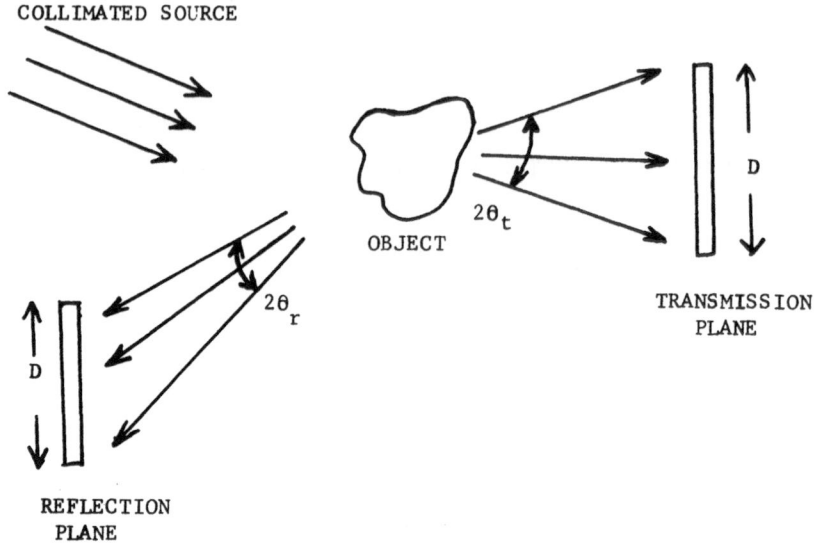

COLLIMATED SOURCE

OBJECT

$2\theta_t$

$2\theta_r$

TRANSMISSION
PLANE

D

REFLECTION
PLANE

D

FIGURE 1

the impedance mismatch between adjacent structures.

Two positions are indicated for the image receptor.
In either case, note that the distance between the object
and the receptor as well as the lateral extent of the receptor, D, will limit the maximum angle of energy reception.
This angle ultimately defines the obtainable resolution.
Consider this heuristic explanation. Choose two closely
spaced points on the receptor and then trace them back a-
long ray paths to one point on the object. Then, trace
back two widely spaced points. The greater angular separation between the two rays results in better "definition"
of the location of the point of intersection on the object,
and hence better resolution.

The resolution of an imaging system is determined by
D and θ_r (or θ_t), assuming, of course, that the angular response of the receptor does not impose a further restric-
tion. The "figure of merit" for resolution is the so=cal-
led numerical aperture, NAm and, by definition, NA=sin θ_r
(for the reflection mode).

Diffraction theory ultimately limits the obtainable resolution to one half of the illumination wave length, L, and
the resulting theorem for lateral resolution, ΔX, contains
both factors L and NA as follows:

$$\Delta X = \frac{L}{2\ NA} \tag{1}$$

Once the resolution is characterized for a system, a depth of focus (or depth of field) ΔY can be determined. Physically, this is the length of the zone, along the principle direction of propagation of the scattered or transmitted illumination, that the image of a point source object retains its apparent size ΔX. Governed by the same factor as above, it is quite easily shown that

$$\Delta Y = \frac{L}{2(NA)^2} \tag{2}$$

Figure 1 illustrates an operating mode which does not employ a lens to focus the image upon the receptor. If the location of the receptor plane is beyond the distance ΔY from a structure of size ΔX in (or on) the object, this image detail will be blurred. However, if a reference beam is directed toward the receptor plane without interacting with the object, a hologram is produced which, if properly reconstructed, reproduces focused images of object structures. The holographic aspects of imaging are discussed elsewhere in this volume

If a lens can be incorporated into the system (this is not always possible or practical, as for example in X-ray imaging), additional modes of operation are possible. Consider the through transmission case for simplicity and refer to Figure 2.

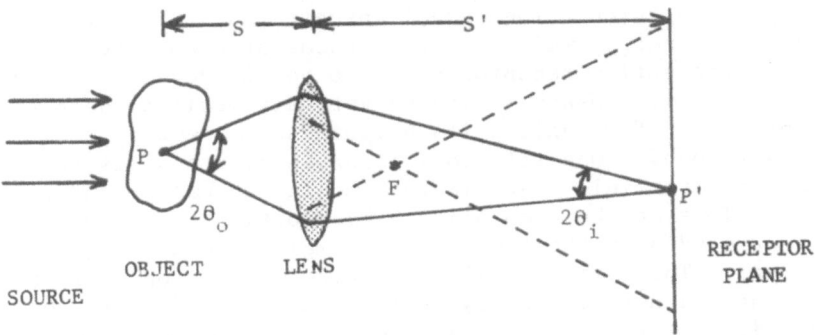

FIGURE 2

Here, the receptor plane is located at a distance S' from the lens in order to correctly image point P of the object which is located at a distance, S, from the lens as shown. The well-known relationship which governs imaging with a simple lens is

$$\frac{1}{S} + \frac{1}{S'} = \frac{1}{F} \qquad (3)$$

where F is the focal length of the lens. Furthermore, there is lateral image magnification, M, determined by the ratio of S and S'; specifically,

$$M \approx \frac{S'}{S} \qquad (4)$$

There are further considerations which will be given to Figure 2. First of all, as the scale of the illustration shows, the angular scatter of energy from P, $2\theta_o$, is demagnified at the receptor plane to some other value, $2\theta_i$. Quantatively, θ_i/θ_o is proportional to $1/M$. The important point, from a practical viewpoint is that the acceptance angle of a typical receptor may be intrinsically limited, thereby producing an effectively lower NA, and hence poorer resolution. This lens approach, which magnifies the image and demagnifies the angular spread of the energy, can be used to advantage "squeeze" more of the angular information into the receptor within its range of response.

A second consideration has to do with a point such as P, which may scatter only a very small fraction of the incident energy. Presuming that the remainder of the energy will pass on through the object unscattered, this, still collimated energy, will come to a focus at F and then diverge, causing the entire receptor plane to be flooded with "background energy." Hence it may be very difficult to detect the presence of P' unless its brightness exceeds that of the background. One solution, shown in Figure 3, is to create a dark field image instead of the normal bright field image. This can be accomplished by blocking the energy concentrated near the focus with a small disc of opaque material. This disc is called a "zero order stop". Ideally, only the unscattered energy is blocked if the diameter of the disc, d, is sufficiently small and points such as P' will be perceived more easily.

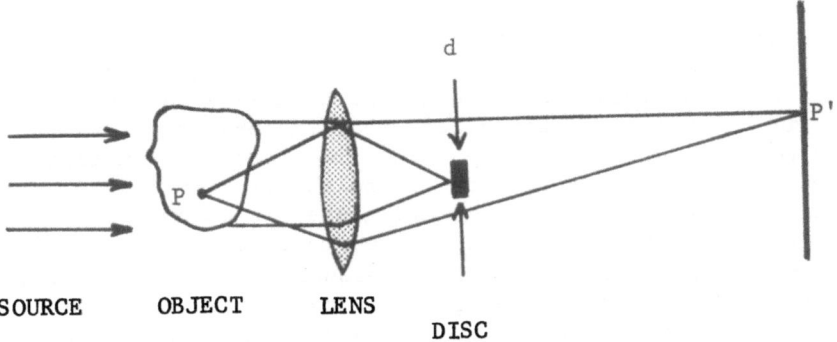

SOURCE   OBJECT   LENS

DISC

FIGURE 3

To describe another mode of operation, assume that the object
is completely transparent to the illumination but that P
causes a small change in the phase of the forward, scattered
energy. If the source is incoherent, the receptor plane in
Figure 2 would be uniformly bright. Although the phase at P'
would be altered, the time averaged intensity of the energy
would be the same as at the adjacent region. Although the
dark field mode in Figure 3 would reveal P', another solution is to replace the opaque disc, shown in Figure 3, with
a transparent disc that shifts the phase of energy passing
through it by 90° with respect to the energy going around
it. This was first done by Zernicke and references to this
"phase contrast" technique may be found in textbooks concerned with optical microscopy. Another solution to this
problem is to use electronic detection means at the receptor plane. Phase detection can be easily accomplished
electronically if the frequency is within the range available instrumentation, such as is the case with acoustic
imaging.

Categories of Acoustic Imaging Systems

Acoustic imaging techniques may be separated into two
distinct classifications which also have separate origins.

In the 1920's, S. Sokolov devised techniques for producing acoustic images in a functional manner analogous to that of an optical camera or vidicon. That is, the object to be visualized is flooded with sound and an area-sensitive, a-coustic to optic receptor placed near the object so as to receive the scattered energy. The image produced in this manner is an orthoscopic view of the object, similar to an ordinary photograph and all imaging systems which present this type of image may be classified together.

Evolving separately were techniques for sonar and radar wherein short bursts of energy were transmitted out direct-ionally into space. By scattering a portion of the energy back to a receiver, objects which lie in the path of the beam could be detected. In this type of system the time of flight of a pulse corresponds to the distance between an object and the receiver and a CRT display of echo amplitude as a function of time represents one line of an image. A two dimensional image may be built up of many lines which are separated by polar or rectilinear geometries. The scanning mechanism for the transmitter-receiver is synchronized with the CRT monitor and the image is a crossectional view rather than ortho-scopic. Although this class of imaging systems is not pri-marily considered in this text, it is a very popular and useful technique as evidenced by the ready availability of commercial apparatus.

Characteristics of Sound Propagation

The image receptor is the least common denominator of ultrasonic visualization systems being considered and em-ployed today. Each not only differs with respect to sen-sitivity and practicability, but, furthermore, rely upon a different type of effect. Therefore, before describing the physical interactions that could be employed for imaging systems, an example of the physical effects which occur when a sound wave propagates is presented. In this example , shown in Table 1, the assumption has been made that a linear wave equation is valid and that a periodic, sinusoidal, traveling wave of frequency 1 MHz is launched into water. By making use of one of the effects that occur, an acoustic imaging system might be constructed.

System Concepts

A review of Table 1 indicates that the passage of a sound wave in water is associated with relatively great

TABLE 1

Physical Properties of Water

Mass density	1.0 gm/cc
Optical ondex of refraction	1.33
Velocity of sound	1500 m/sec

Environmental Conditions

Atmospheric pressure	1 atm
Water temperature	30° C

Acoutic Wave Parameters

Acoustic frequency	1 MHz
Intensity level	1 W/cm^2
Radiation force upon a reflector	134 dynes/cm^2

Peak Instantaneous Values of:

Pressure	1.7 atm
Particle displacement	183 Å
Particle velocity	11.5 cm/sec
Strain wave	7.62×10^{-5}
Temperature wave	3.8×10^{-3} C°
Optical index of refraction	2.8×10^{-5}
Density	7.62×10^{-5} gm/cm^3

fluctions in localized pressure amplitude, compared to any
of the other parameters. Therefore, it might be expected
that a predominantly pressure (or force) sensitive, acoustic
detection process would be the most sensitive type of receptor. That being the case (as shown in Chapter 4) the
first type of image receptor to be considered is the, intrinsically pressure sensitive, piezoelectric type of systems.

a) Piezoelectric systems

The piezoelectric effect, first discovered by Pierre
and Jacque Currie in the 1880's, involves the development

of an electrical charge upon the surface of certain materials in response to an applied force. If the material is
unelectroded, the charge remains confined to the area of the
applied force and the distribution of charge is a replica
of the sound field. The "Sokolov tube" is based upon this
effect.[1] Referring to Figure 4, acoustic energy scattered
from an insonified object is incident upon a piezoelectric
element which forms the front plate of a vidicon type of
tube. The rear face of the piezoelectric is unelectroded,
thereby developing the charge replica of the sound field.
In order to read the charge, a high velocity electron beam
is scanned across the back surface point by point in a TV
rastor fashion. Secondary electrons emitted in the process
are modulated by the charge distribution and collected by
the electron multiplier. The resulting signal is amplified
and fed to a TV monitor which is synchronized to the image
converter tube.

The resolution capacity of the Sokolov tube is governed by the velocity of sound difference between the piezo
receptor plate and the sound conducting medium in front of
it. For a quartz and water interface, the effective half-
angular aperture would be less than the critical angle of

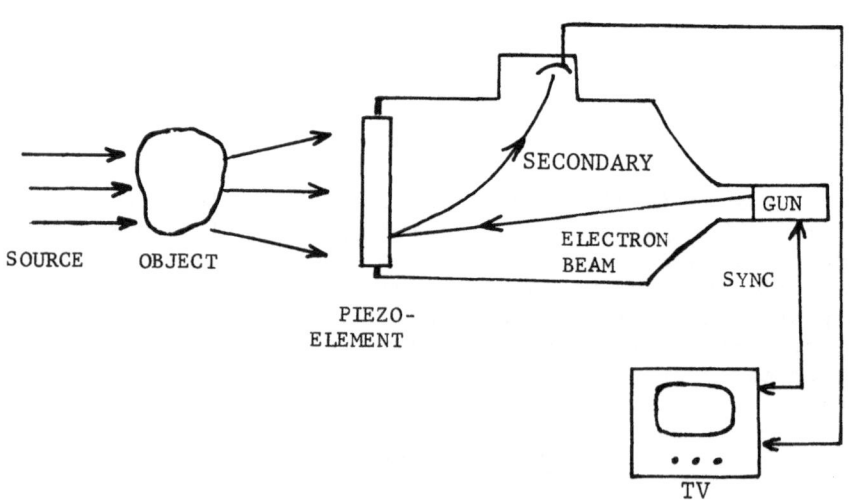

FIGURE 4

14°, say 10°, therefore, NA = 0.173 and Δ X = 4.5 mm at 1 MHz.

A simpler system for piezoelectric reception is a single element, microphone, which is mechnically scanned over the receptor plane. Here, energy is more directly transferred from mechanical to electrical forms than in the Sokolov tube. The effective element size though must be physically smaller than the expected resolution dimension which is determined by the wavelength of the sound and the NA of the system lens.

Elaborating on direct piezoelectric systems instead of mechanically scanning a single element, a two dimensional array of elements, fixed in space, can be employed. In different embodiments of this system individual elements can be sequentially switched into the receiving electronics or, instead, elements can be operated simultanenously through multiple isolated channels. This is discussed further in a later chapter.

One last discussion pertaining to arrays is in the context of the pulse-echo (non-orthoscopic) type of system for imaging. Recall that in order to produce a two dimensional image, separate spatial positions of the beam, are necessary. With arrays of transducers operating in the transmit and/or receive mode, the sound beam can be electronically steered. This is accomplished by appropriately phasing the energy to the elements. Furthermore, with appropriate two dimensional phasing, electronic focusing may be accomplished.

An interesting type of piezoelectric receptor has been proposed by Auld and others which is based upon an optical-equivalent to the Sokolov tube[2]. Referring to Figure 5, the receptor element here is a material which has both piezoelectric and photoconductive properties. These properties interact, however, in such a way that in the presence of optical illumination the piezoelectric effect is quenched. The information output is taken from across the electrodes of the element and contains the spatially averaged sound signal distribution everywhere except the region that is illuminated by the laser. The image is thus "negative" instead of "positive", however, this can be reversed electronically. There are interesting variations of this system which are worth mentioning. By illuminating the rear of the receptor with a two dimensional optical image of, say, a Fresnel zone plate, the receptor can be self focusing.

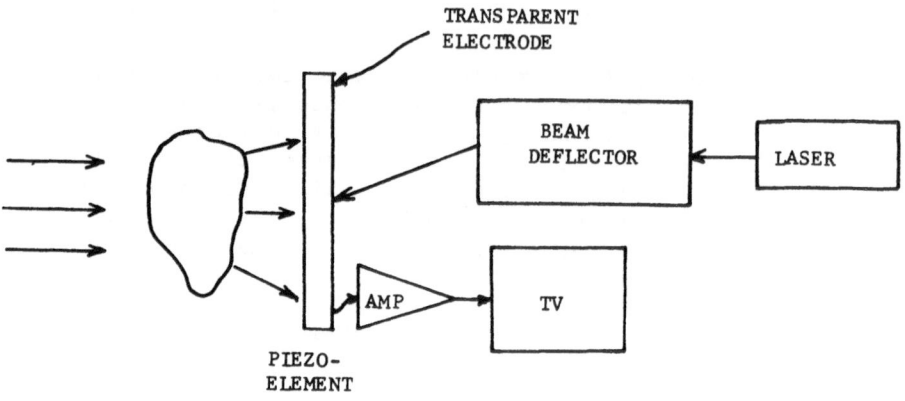

FIGURE 5

Furthermore, by varying the size of this optical image and by translating the position of the image, the position of the focus may be scanned in all three dimensions.

b) Thin layer receptors

In a great variety of ways the action of ultrasonic energy upon an optical property of a window type of device may be employed for direct acoustic to optical translation. Figure 6 schematically shows the simplicity of such a viewing system for the case in which the optical reflectivity of the receptor changes in response to the sound. In other embodiments of this type of system optical density change effects and polarization shifting mechanisms require slight modifications of the illustrated diagram.

A few particular examples of image receptors of this type are here. The "Pohlman Cell"[3] consists of a fluid suspension of small metal flakes combined between two parallel plates, one of which is acoustically transparent: The particles are nominally disc shaped and are optically reflective. The hydrodynamic forces associated with the sound field (entering through the acoustic window) tend to align the particles such that their face normals are

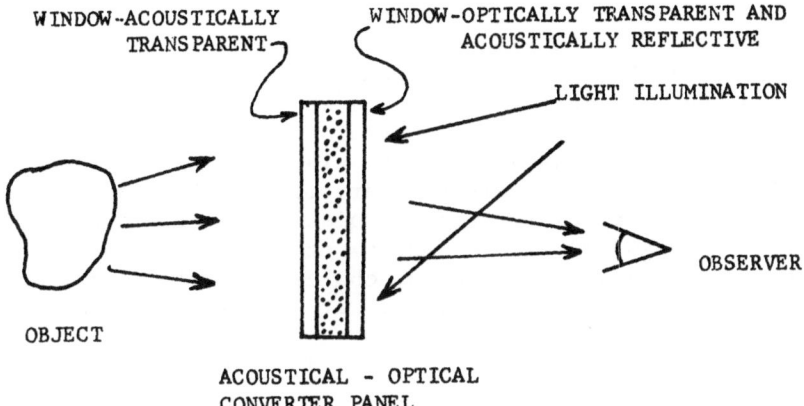

FIGURE 6

parallel to the sound field normal. Without the sound field, the particles assume a rather random orientation. It has been shown, that instead of assymetric particles in suspension, small diameter latex spheres will also work (Cunningham and Quate)[4] although the physical mechanism may be somewhat different. In particular, the acoustic radiation forces can cause the local particle population level to vary according to the sound distribution. Some of these migration effects also occur in the Pohlman cell configuration.

There are two types of liquid crystal materials which can be used to transfer acoustic wave information into a visual display. The first, and common type of liquid crystal, of the cholestric type, changes color in response to temperature. Examination of Table 1, however, indicates that the temperature fluctuations are very small in amplitude. Furthermore, the response time of liquid crystals may be too long to respond to megahertz frequencies. On the other hand, if the acoustic energy is absorbed by a material and thereby turned into heat, then a liquid crystal layer may be employed as a temperature sensitive panel, whose optically revealed pattern is representative of the acoustic field.[5]

The second type of liquid crystal that can be ultra-

sonically stimulated is the nematic type. This type of
material can impart a change in polarization of through
transmitted light, or instead, cause a significant change
in the amount of light scattered; that is, the material can
turn from a transparent to cloudy state. Typically these
effects are induced by applying an electric field across a
thin layer of the material, however, it was discovered a
few years ago that ultrasonic stimulation is also effective.
(Kessler and Sawyer) The mechanism of action of nematic
crystals is more direct than for cholestric and depends upon the hydrodynamic forces allied with the sound energy.

c)   Surface distortion methods of reception

The passage of a sound beam is associated with minute
displacements of the particles that constitute the propagation medium. Table 1 indicates a value for displacement
which is far smaller than a typical optical wavelength.
(For red light propagating in air, the wavelength is 6000 Å)
At lower values of intensity or at higher acoustic frequencies the displacements are even smaller. For example at
either 1 watt/cm^2 and 1 GHz or at 1 microwatt/cm^2 and 1 MHz
the displacement amplitude is only 0.18Å, which is smaller
than interatomic spacings. Considering the above, it is not
likely that the distortion which a sound wave imparts to an
optically reflective surface will be readily visible to
the naked eye. For one thing, the pattern is actually an
optical phase replica of the sound field and secondly, the
human eye is not sensitive to optical phase. However, referring back to the general discussion of imaging modes,
optical phase pertubations may be made visible by means of
dark field or phase contrast techniques.

Figure 7 illustrates a sound beam directed towards a
liquid-air interface and the resulting distortion that occurs in this case, first of all, as mentioned above, the normal displacement of the sound beam results in a dynamic corrugation due to the displacement amplitude of the wave, y.
The periodicity of the corrugation is determined by the angle of the imcident sound with respect to the normal to the
surface, and the apparent phase velocity of this interfacial
disturbance,v, is greater than the velocity of sound in
the water, c_o, by the factor $(\sin \theta_i)^{-1}$. The second effect
has to do with force of radiation pressure upon the surface
of the water. Radiation pressure can exert steady forces on
interfaces between media having different values of acoustic

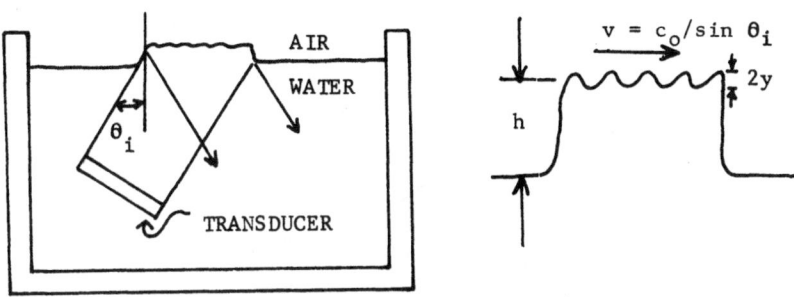

FIGURE 7

impedance and in this case, it causes a static levitation of
the surface, h. The height of the levitation is dependent
upon restoring forces such as gravity and surface tension
of the liquid. However, the static effect is typically
greater in magnitude than the dynamic effect for a liquid-
air interface.

Figure 8 shows a simple imaging system, due to Sokolov
which utilizes a dark field optical system to detect surface
levitation. An improvement to this type of system, developed
by Brenden and others utilizes an acoustic reference beam
as shown in Figure 9.[7] The reference beam, which construct-
ively interferes with the object beam results in a static, but
periodic corrugation to the surface which is significantly
greater in amplitude than the dynamic ripple, that is, for
acoustic frequencies in the few megahertz range. These
static corrugations help to relieve some of the constraints
of the optical viewing system such as the case in Figure 8,
where most of the light is concentrated too close to the
zero order stop. Furthermore, the efficiency of light diffraction is improved by the increased depth of the static
corrugation.

Although Figures 8 and 9 illustrate methods of visual-

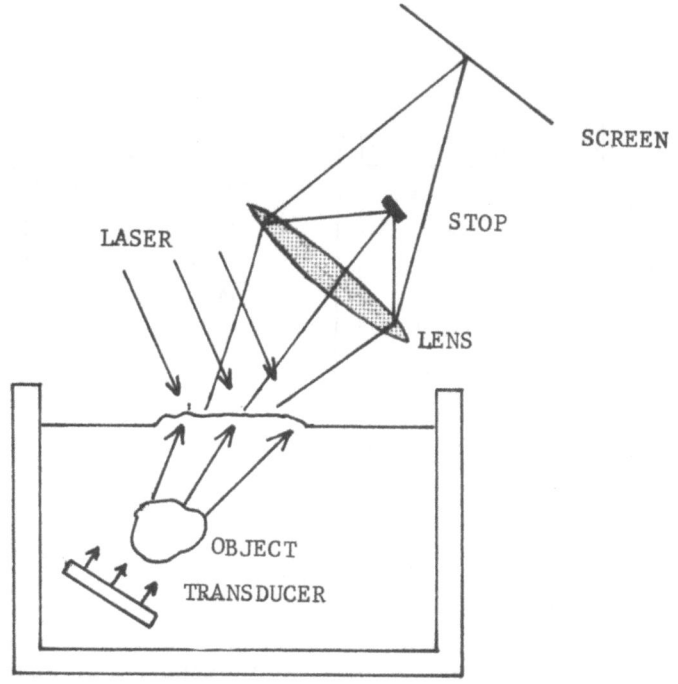

FIGURE 8

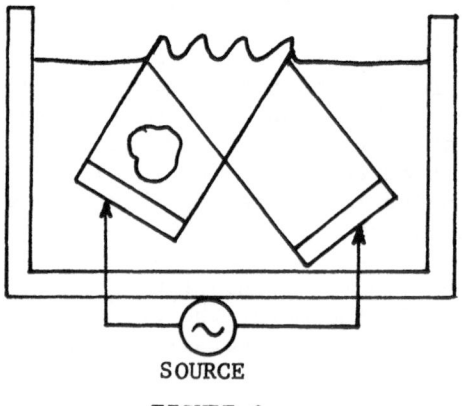

FIGURE 9

ization which permit direct optical translation of the a-
coustic image, there are several alternative approaches
which have improved optical reflectivity at the interface
and rejection of stray light which may "wash out" low intensity images. These alternative approaches employ opto-
electronic methods to detect distortion of the surface, and
in general, these techniques require point by point scanning
of a laser beam. An attractive aspect of this type of system is that it is considerably easier to rapidly scan a
laser beam than it is to mechanically scan a receiving
transducer.

In the first approach to be described, an optical interferometer may be set up to measure the phase shift between two coherent light beams. Neglecting scanning aspects
for the moment, consider the two separate paths taken by the
laser light before arriving at the photodetector in Figure
10. In one path, the phase of the light is shifting continuously by virtue of its having been reflected from a
dynamically rippled surface due to the sound. The second
path is only statically phase shifted from the stationary
reference mirror. However, from the example shown in Table
1, the magnitude of the dynamic phase shift is so small that

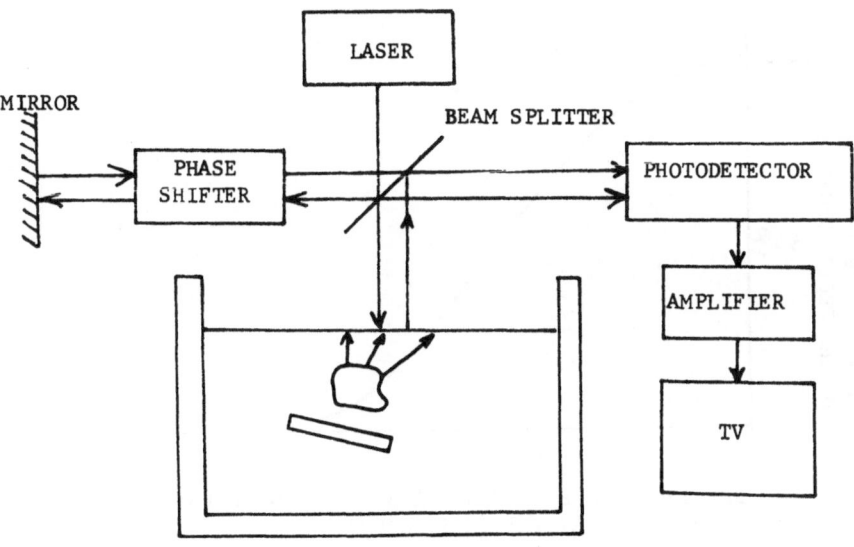

FIGURE 10

total constructive and destructive interference cannot both occur as the surface moves up and down. Figure 11 shows, quantitatively, how the light intensity falling upon the photodetector varies as a function of the phase difference between the two beams. Note that, the sensitivity of the system, i.e., the change in brightness per degree of phase oscillation is very bad near 0 and π and optimum near $\pi/2$. Therefore, in order to insure good sensitivity with this technique an optical phase shifter is inserted in one leg of the interferometer to bias the system near optimum on the sensitivity curve. However, since interferometers are so intrinsically sensitive to vibration and to irregularities of the reflecting surfaces, it is extremely difficult to maintain the correct bias point. However, significant progress has been made by Mezrich et al[8] who have inserted a continuously sweeping optical phase shifter into the reference leg of the interferometer. This device purposely drives the bias point of the system through its optimum point sometime during the sweep. Then by electronically detecting the peaks of the signals, even though some signal to noise ratio degradation occurs, the interferometry concept can be harnessed for acoustic imaging. In illustrating Figure 10, the sound was directed at an air-water interface

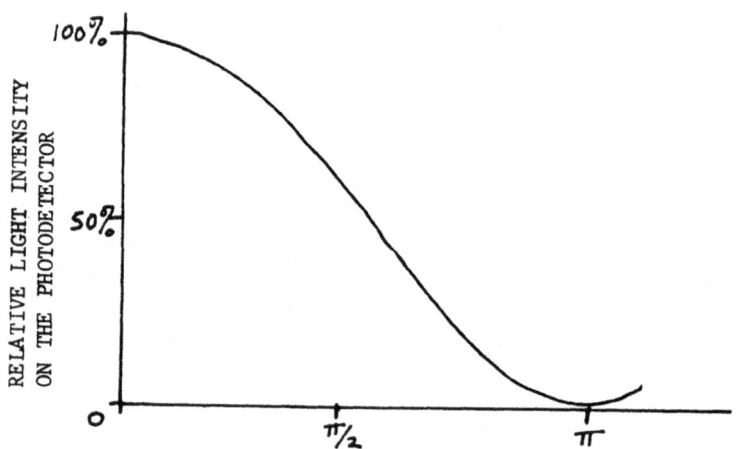

RELATIVE PHASE DIFFERENCE BETWEEN TWO EQUAL INTENSITY LIGHT BEAMS AT THE PHOTODETECTOR

FIGURE 11

similar to that in previous figures. This need not be the
case. In fact, Mezrich actually employs a thin acoustically
transparent, but optically reflective, pellicle which is totally submerged in the water. Others employ a liquid-solid
interface in which the solid is acoustically absorbing but
optically transparent, except for an optically reflective
coating at the interface.

Another method for detecting the sound amplitude at an
interface relies upon slope variations of an acoustically
perturbed surface. Recall from Table 1 that a 1 watt/cm^2 beam
at 1 MHz in water causes particle displacement of 183 angstroms. If this wave is incident at 45° to an acoustically
matched but optically reflective surface, the normal component of displacement is 130 angstroms. The acoustic wavelength of 15 micrometers is projected into the surface as
21 micrometers and, roughly speaking, the maximum localized
slope angle of the surface, from flatness, is 1 milliradian.

Figure 12 illustrates the so-called knife edge method
for detecting surface slope. A focused laser beam is directed to the receptor boundary. The angular position of the
reflected light is then modulated by the dynamic activity
of the corrugations. The photodiode, which responds to light

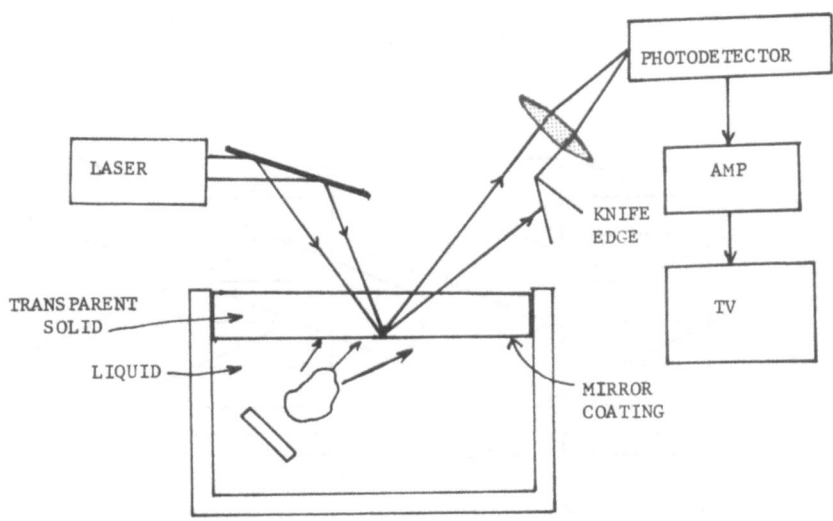

FIGURE 12

intensity fluctuations is not intrinsically sensitive to
angular modulation of the light. However, an appropriately
positioned obstacle such as a knife edge, will block a
fraction of the light which depends upon the instantaneous
angular position of the beam. Thus, the knife edge demodulates
the angular position into corresponding intensity fluctuations. This system is very simple and has been employed by
this author and colleagues for acoustic imaging over the
frequency range 1-500 MHz so far.[9]

d) Light-Sound interaction

It is well established that a sound wave propagating
in an optically transparent medium will act as a diffraction
grating for a light beam[10] This effect, which is now commercially employed to scan and modulate laser beams, may also
be employed for acoustic visualization. The diffraction
grating effect is caused by acoustically induced modifications to the optical index of refraction. If we assume that
the so-called Bragg condition is valid, it is quite simple
to develop a basic concept for an imaging system. The
Bragg condition states that if a ray of incident light interacts with a ray of sound at an angle $\pi/2 \pm \emptyset_B$ then the light
will be scattered from its original direction by an angle
2 $\emptyset_B$ where $\emptyset_B$ is the Bragg angle and is equal to $\lambda/2\Lambda$.

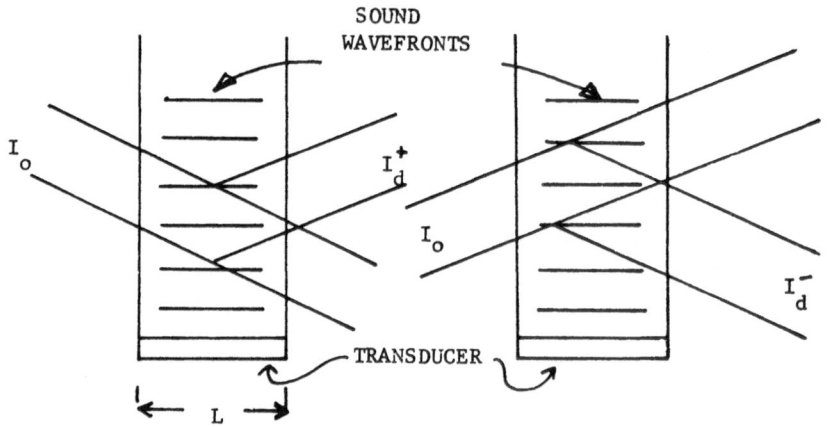

FIGURE 13

Here, $\wedge$ is the optical wavelength and $\wedge$ is the acoustic wavelength.

Figure 13 shows a collimated beam of light interacting with a collimated beam of sound at the 2 possible Bragg angles. The acoustic wavefronts seem to act as simple mirrors for the sound. Depending upon the interaction length, L, and the acoustic pressure level (and therefore, the index of refraction pertubation) a fraction of the incident light, I_o, will be diffracted, I_d, and the fraction can approach 100%. In addition to changing direction of the light, this diffraction process also shifts the frequency of the incident light beam by an amount equal to the sound frequency. This is due to the fact that the "diffraction grating" is a traveling wave rather than being stationary in space. The light frequency is upshifted in 13a and downshifted in 13b.

Figure 14 shows the diffraction which occurs when a line source of light, L, interacts with two sources of sound, O_1 and O_2.[11] Because of the variety of rays available for interaction, both upshifted and downshifted orders occur, in general. Furthermore, the 2 sources of sound have diverted light, originally headed for L, to points O_1^+, O_1^-, O_2^+, and O_2^- each of which represents a light field replica of the sound sources, O_1 and O_2. If drawn to scale, it would become more evident that the distance between O_1^+ and O_2^+ or between O_1^- and O_2^- is demagnified from the original spacing ΔX by the ratio of the light and sound wavelengths. At 1 MHz for example, using red light the image is reduced by a factor of about 2000. However, this demagnification is only in the plane of the paper. In the other direction, perpendicular to the paper, there is no demagnification. Therefore, to correct this astigmatic imaging process, a cylindrical optical system is required.

Concluding Remarks

A rather simplistic overview of systems for acoustic visualization was presented in this chapter in order to introduce some of the concepts that are discussed more thoroughly in later chapters. No attempt has been made to present all types of imaging systems being considered. Rather a few examples which are relatively easy to explain have been chosen. Ultimately, the great variety of systems must be brought together over a common denominator for

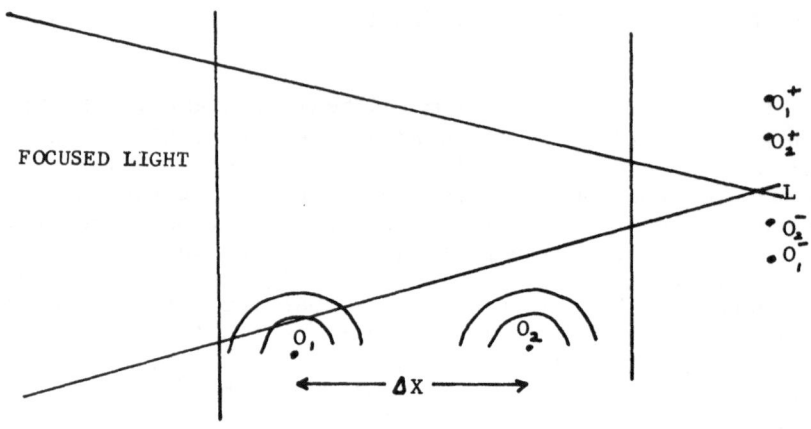

FIGURE 14

comparison. Actually there are two such denominators,
overall sensitivity and practicability. Unfortunately, the
latter is continuously undergoing thorough revision and
these aspects will not be discussed here. On the other
hand, the theoretical basis for limitations on sensitivity
is worked out for a number of systems and it is the subject
of the next chapter.

REFERENCES

1) a. S. Sokolov, USSR Patent No. 49(August 1936), British
      Patent No. 477139(1937), and U.S. Patent No. 21 64
      125(1939).

   b. J.E. Jacobs, "Ultrasonic Image Converter Systems
      Utilizing Electron-Scanning Techniques," IEEE
      Trans. Sonics Ultrason. SU-15 146-152(1968).

2) a. B.A. Auld, R.J. Gilbert, K. Hyllested, C.G. Roberts,
      and D.C. Webb, "A 1.1 GHz Scanned Acoustic Micro-
      scope," in Acoustical Holography, edited by G.Wade
      (Plenum. N.Y., 1972), Vol.4, pp. 73-96.

b. K. Wang, H. Shen and G. Wade, "An Opto-Acoustic
    Transducer for High-Sensitivity Ultrasonic Imaging"
    Proc. 1975 Ultrasonics Sumposium Cat. No. 75-CHO-
    994-4SU. IEEE New York, NY 10017

3) R. Pohlman, "Material Illumination by Means of Acoustic-
    Optical Imagery," Z. Phys. 1133 697(1939). See also Z.
    Angew. Phys. 1, 181(1948).

4) J.A. Cunningham and C.F. Quate, "Acoustic Interference
    in Solids and Holographic Imaging," in Acoustical
    Holography, edited by G. Wade(Plenum, N.Y., 1972), Vol 45,
    pp. 51-71.

5) Bill D. Cook, University of Houston, Dept. of Mechanical
    Engineering, Unpublished

6) L.W. Kessler and S.P. Sawyer, "Ultrasonic Stimulation of
    Optical Scattering Nematic Liquid Crystals," Appl. Phys.
    Lett. 17, 440-441(1970).

7) B.B. Brenden, Chapter 6, this book.

8) R.S. Mezrich et al., "Ultrasonovision", Proc. 1974
    Ultrasonics Symposium, Cat. No. 74-CHO-896-SU, IEEE, Inc.,
    N.Y., N.Y. 10017

9) a. L.W. Kessler, "Review of Progress and Applications
        in Acoustic Microscopy.", J. Acoust. Soc. Amer. 55,
        909(1974).

    b. R.L. Whitman, M. Ahmed and A. Korpel, "A Progress
        Report on the Laser Scanned Acoustic Camera",
        Acoustical holography, Vol. 4, ed. by G. Wade,
        Plenum Press, N.Y.(1972).

10) R. Adler, "Interaction Between Light and Sound", IEEE
    Spectrum, pp. 42-54, May 1967, IEEE INC., NY 10017.

11) a. A. Korpel, "Visualization of the Cross Section of
        a Sound Beam by Bragg Diffraction of Light," Appl.
        Phys. Lett. 9, 425(1966).

    b. H.V. Hance, J.K. Parks, and C.S. Tsai, "Optical
        Imaging of a Complex Ultrasonic Field by Diffraction
        of a Laser Beam," J. Appl. Phys. 38, 1981-1983(1967).

Chapter 4

SPECKLE AND SENSITIVITY IN MODERN SYSTEMS

G. Wade

Electrical Engineering Department

University of California at Santa Barbara

I. INTRODUCTION

In the previous chapter, Dr. Kessler has shown that it is possible to categorize the many types of modern acoustic imaging systems in various ways. One simple way, which we shall use below, is particularly clear cut in terms of ease of categorization. In addition it is especially effective in distinguishing between systems on the basis of the very important behavioral characteristics associated with speckle and sensitivity.

We divide the systems into two groups. The first group consists of all systems employing laser beams to read out the image information. The second group comprises those systems which use piezoelectric elements for image readout. This chapter will describe the behavior of these two categories of systems in terms of inherent sensitivity and freedom from certain spurious image elements called speckle and ringing. Obviously in such applications as medical diagnosis high sensitivity can be very important in order to avoid tissue damage by having to use too high a level of insonification. In almost all imaging situations, speckle and ringing are unwanted. They are particularly detrimental whenever fine detail of an object is desired.

Thus both sensitivity and absence of speckle and ringing are important characteristics for high-quality imaging. As we shall see from the arguments presented below, the category of systems using piezoelectric readout

is inherently superior to the category employing laser-
beam readout as far as these characteristics are concerned.
This will be discussed here only in qualitative terms. The
rigorous mathematical bases which permit quantitative
determinations in this regard will not be presented, but
can be found in the references to the research literature
which are cited in this chapter.

II.   SPECKLE AND RINGING

Let us first consider the question of speckle and
ringing. As is well known, any system that is fundamentally holographic in nature uses coherent waves and operates in such a way as to retain the phase information
associated with these waves. On the other hand, in most
non-holographic systems, this information is not preserved.
Generally speaking, phase is preserved in the laser systems
while it is not preserved in the piezoelectric systems.
Retaining the phase makes a fundamental difference as far
as the presence of such interference products as speckle
and ringing are concerned.

Precisely what is meant by speckle and ringing?
Speckle refers to randomly positioned variations in image
intensity due to phase cancellation and phase reinforcement among all the coherent spatial wave components
impinging upon the image plane. Speckle is therefore
characterized by randomness. Ringing, on the other hand,
is more regular in appearance. Ringing refers to systematic variations, also due to phase cancellation and phase
reinforcement, which show up in the image as fringes.
Ringing can be thought of as a more orderly manifestation
than speckle of the interference phenemenon. Usually the
fringes due to ringing are particularly prominent in the
vicinity of the images of sharp edges. Thus the coherent
spatial wave components used in forming an image can and
do interfere with each other to produce interference products that would not be present if the components were
incoherent.

In ordinary sight we use incoherent light and do not
have to cope with speckle or ringing. We are therefore not
used to seeing interference fringes and spots in the images
formed on the retinas of our eyes. When we observe speckle
and ringing in images produced by coherent systems, it is

difficult to interpret what we are seeing. These spurious
elements are not the same as noise. They actually represent information, but a type of information we normally do
not use. If we were only clever enough, perhaps we could
garner meaning from them. However, because of our lack of
ability to interpret what we are seeing, they usually
detract from, rather than add to, our understanding of the
object.

Speckle and ringing are particularly annoying in still-
life pictures. They seem to go away if the image is moving.
The reason for this is physiological. When the eye looks at
a moving image produced by a real-time holographic system,
the spurious elements change shape and shift position rapidly. A visual averaging or integrating effect takes place
that, from the standpoint of the observer, seems to smooth
out the deviations. Thus in a moving image, an observer will
often not find speckle and ringing to be particularly dis-
turbing, and may actually be quite unaware that the spurious
elements are even present. However, if a photograph is
taken of the image at some instant of time, the motion, and
hence the averaging, are eliminated and the interference
products do indeed become apparent.

The problem of ringing and speckle in such a photograph
is illustrated in Fig. 1. The system in this case is one
that employs Bragg diffraction of laser light from coherent
sound in order to produce an optical replica of an acoustic
beam which has been scattered from the object. By processing
this laser light with a conventional lens system, the image
is obtained. Chapter 9 describes Bragg-diffraction imaging
in detail. For the purposes of the present discussion, it
is sufficient to point out that when light is Bragg-diffracted from sound, the phase information associated with the
sound is retained in the diffracted light. The images from
a Bragg-diffraction system, therefore, usually contain much
ringing and speckle. The ringing is especially prominent
in Fig. 1 and shows up mainly as a number of sets of parallel
fringes associated with various edges in the object or in the
system itself.

One solution to the problem of ringing and speckle is
to use non-coherent waves rather than coherent waves. This
fact was illustrated nicely by some experiments performed a
few years ago on a Bragg-diffraction imaging system, in

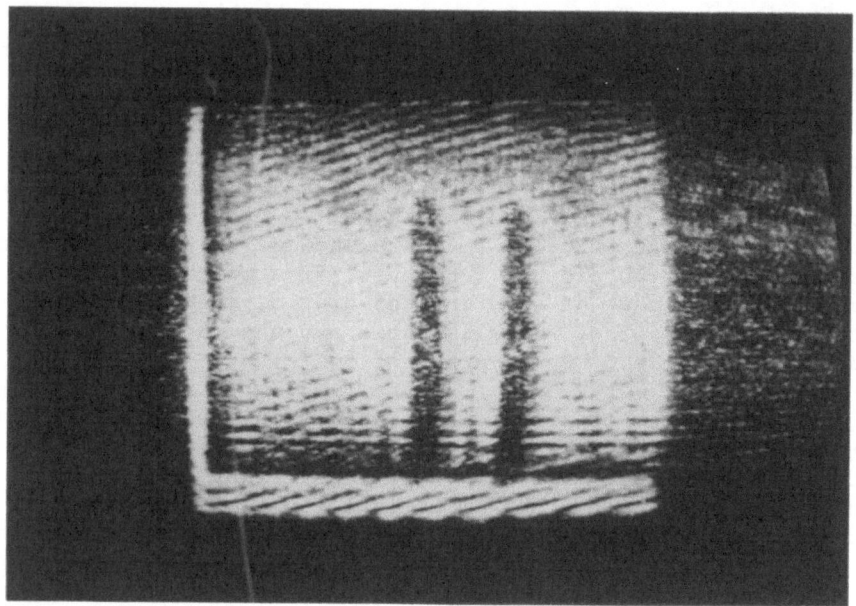

Fig. 1 Ringing and speckle in a photograph of an image
        from a Bragg-diffraction imaging system. The
        object was an aluminum plate with two holes
        drilled into its lower edge. The ringing is
        particularly noticeable appearing in the form of
        several sets of parallel fringes.

which a transducer of relatively large active area was
needed [1]. In building the system it would have been
desirable for simplicity to have been able to use a single,
large transducer for producing the ultrasonic beam. However,
considerations of cost and the likelihood of breakage were
important factors in the decision to form the beam with a
mosaic of six small transducers, rather than a single large
transducer. Once the system was made operational, a direct
image of the acoustic beam emerging from the transducer array (with no object having been placed in the specimen tank)
was photographed and is seen in Fig. 2. Prominent inhomogeneities were obviously present. This at first was attributed to deficiencies associated with the transducer array.
However, it was soon noted that slight changes (± 5 KHz) in
the r.f. drive frequency to the transducers resulted in
large shifts in the position and shape of the intensity
variations throughout the image. This fact suggested that
the inhomogeneities were due to speckle. A practical solution to the problem was immediately apparent. The variations
could be smoothed out by simply introducing frequency modulation at the signal generator for the transducer (to reduce
coherence) with a total swing of about 10 KHz. A modulation
frequency of 1 KHz was used, and the resulting image, considerably smoother, is shown in Fig. 3.

This was a relatively simple solution to a difficult
problem and seemed to offer no disadvantages provided that
the frequency deviation was not too high. For deviations
as great as 50 KHz, it was noted that motion in the position
of the image due to the acoustic frequency change was sufficient to blur the image detail.

As stated above, the source of the inhomogeneities was
first thought to be transducer deficiencies. However, after
sophisticated experiments were performed, it was discovered
that the inhomogeneities were almost entirely associated
with multiple acoustic reflections within the cell and
hence were simply a severe manifestation of interference
between the various sets of coherent waves propagating
there. Thus it was principally a form of speckle that was
being observed. Substantial improvement could be effected
by simply placing sufficient acoustic absorbent material
in strategic places in the cell. Evidently transducer
deficiencies, such as lack of parallelism between the elements of the transducer array, were not particularly critical because none of the inhomogeneities seen in Fig. 2

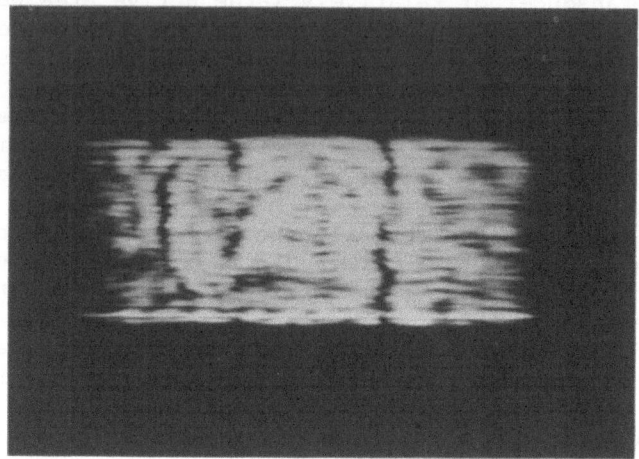

Fig. 2   Image of an acoustic beam energing from a trans-
         ducer array in a Bragg-diffraction imaging system.
         Note the prominent inhomogenities in the image
         brightness pattern.

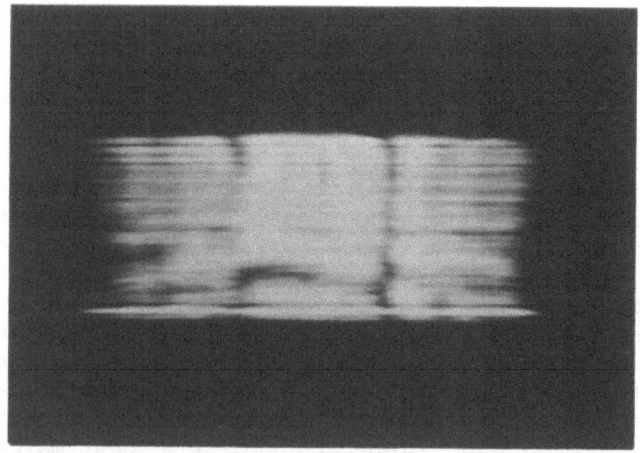

Fig. 3   Image of the same acoustic beam as in Fig. 2 when
         the transducers were frequency modulated to reduce
         speckle.

could be directly attributed to this type of deficiency. Apparently, as described above, the inhomogeneities were purely a matter of speckle being produced by multiple acoustic reflections within the cell. Because of the long path lengths associated with the reflections, a slight change in frequency could produce a large change in the phase of the reflected waves and hence a large change in the positions of regions of reinforcement and cancellation in the image. Thus, for this case, the inhomogeneities could be smoothed out by a slight reduction in the coherency of the waves brought about by the frequency modulation.

The above conclusion is consistent with the observation that the problem of speckle and ringing is greatest in the systems using laser-beam readout where the need for coherence in the acoustic waves is also the greatest. The severity of the problem in these systems can be reduced by decreasing the coherence of the waves being used. However, generally in these systems there is a limit to how much decrease in coherence can be tolerated because image blurring will eventually occur. Blurring is a common consequence of reducing coherence since resolution decreases with decreasing coherence.

The use of frequency sweeping to eliminate this type of spurious image detail in the general case of acoustic imaging with monophonic (single frequency) insonification, has been put on a firm mathematical basis by Korpel, Whitman and Ahmed [2]. In addition to frequency sweeping, they have suggested other ways to produce the necessary temporal variations, such as using noise to drive the insonifying transducer or moving a scatter plate through the insonifying sound field. This problem is also commented upon by Professor Quate in Chapter 12 of this book.

III. SENSITIVITY

As previously mentioned, sensitivity is another characteristic in which piezoelectric readout is inherently superior to laser-beam readout. For such applications as medical diagnosis, where low-level insonification is necessary to avoid damaging the organs and tissue of the patient being diagnosed, high sensitivity is extremely important.

Laser-beam systems are relatively insensitive because the readout process is inherently noisy. Laser beams, by themselves, are plagued with high quantum noise. If it were possible to obtain a noiseless laser beam, the laser-beam systems would have about the same sensitivity as the piezoelectric systems.

In any system the only real limitation to the detection of image information from weak signals is the inevitable existence of noise. Obviously no amount of amplification can make it possible to derive information from a signal whose power is substantially below the level of the noise entering the receiver.

For the piezoelectric systems, the type of noise of greatest fundamental importance is the so-called "thermal noise." The amount of thermal noise corresponding to a bandwidth Δf is given by the well-known equation

$$P_{NT} = kT\Delta f \tag{1}$$

where k is Boltzmann's constant and T is the absolute temperature in degrees Kelvin. For the laser systems, the most important source of noise is called "quantum noise." This type of noise is fundamentally different from thermal noise. It is produced by the quantum limit on the possible accuracy of measurements involving an electromagnetic field. A quantum-noise equivalent to Eq. (1) can be written:

$$P_{NQ} = hf\Delta f \tag{2}$$

where h is Planck's constant and f is the frequency of the laser beam.

As indicated above, quantum noise originates in the process of detecting an electromagnetic field because there is a basic quantum limit on the precision of the measurements that can be made of such a field. Quantum noise may ultimately manifest itself as shot noise in phototubes and photodiodes. However, its origin can be traced to electromagnetic field quantization. It is therefore a quantum phenomenon of fundamental nature.

One way to understand the basis for quantum noise is to consider Heisenberg's enunciation of quantum-mechanical uncertainty as it applies to an electromagnetic field. Such a field can not be measured with arbitrary precision owing to its quantum nature. This inability to measure

with complete accuracy can be thought of in terms of an
equivalent noise source. Thus, the uncertainty inherent
in such measurements is attributed to the presence of an
equivalent noise source giving rise to the quantum noise.

As previously noted, quantum noise is fundamentally
different from thermal noise. For example, Eq. (1) shows
that thermal noise, in principle, can be eliminated simply
by cooling all the elements of the system to absolute zero
temperature. On the other hand, Eq. (2) shows that no
amount of cooling will produce any reduction whatsoever in
quantum noise; nor will any other type of processing per-
mitted by nature.

One way to illustrate the nature of quantum noise is
to consider an automobile odometer which reads only to the
nearest mile. Suppose we wish to know how far the car has
traveled from a given starting point. Obviously with such
an odometer we can determine this distance only with limited
accuracy, that is to the nearest mile. The relative importance of the inaccuracy of this measurement, of course,
depends upon how far the car has traveled. For a distance
of five miles, the inaccuracy can be almost 20%; for fifty
miles, only 2%. By applying this type of thinking to the
case of quantum noise, we can conclude that quantum noise
is not so important, in a relative sense, when we are
dealing with very large numbers of quanta. Thus, in our
laser-beam systems, the higher the laser power, the less
important will be the quantum noise.

As indicated above, quantum noise can not be accounted
for by any classical theory. Classical theory always permits field measurements to be made, at least in principle,
with absolute precision. Quantum noise is therefore not
inherent in classical em theory. All noise predicted
classically can be traced to physical noise sources.

Note from Eq. (2) that quantum noise power is directly
proportional to frequency. Because laser-beam photons are
at high (optical) frequencies, the laser beam itself is
extremely noisy. In fact, quantum noise in the laser beam
constitutes the major source of unremovable noise in all
the systems using laser-beam readout. This quantum noise
makes it extremely difficult for these systems to detect
small differences in the intensity of the various sound
components which are scattered by an object. Therefore

if two adjacent object elements scatter sound with only
slight dissimilarities, it will be very hard to distinguish
the two elements from each other in the final image if we
use laser-beam readout. This is true even if the laser
power level is relatively high [3].

We can easily calculate the equivalent noise temperature of the readout process by comparing Eqs. (2) and (3).
The question we ask in doing so is the following one:
What is the equivalent thermal temperature needed to
produce the same noise output as exists in a laser beam
because of its quantum noise? This temperature is computed
by equating Eq. (1) to Eq. (2) and solving for T. When
we do this we obtain

$$T = \frac{hf}{k} \tag{3}$$

If we enter Eq. (3) with the values of Planck's constant,
Boltzmann's constant, and the frequency of the laser beam,
we compute a temperature somewhat in excess of 20,000°K.
Thus the equivalent noise temperature associated with the
readout process alone in the case of the laser systems
amounts to over 20,000°K.

On the other hand, as previously mentioned, the chief
source of unremovable noise in the piezoelectric systems is
thermal noise. Under the usual operating conditions, these
systems function at room temperature and therefore have an
equivalent noise temperature of only about 300°K.

From the above calculations, we might guess that for
ordinary circumstances the ultimate sensitivity of the
laser systems will be almost two orders of magnitude worse
than that of the piezoelectric systems. Recent analytical
study has confirmed this expectation [3] even when we
assume that the laser beams being used are relatively high
powered. However, the ultimate sensitivity referred to
above is still a far cry from the sensitivities actually
achieved in practice at the present time. In fact, the
ideal performance hypothesized in the referenced analytical
study has not yet been approached in any of the systems.
The difference in the sensitivities actually achieved to
date (1975) is even greater than the two orders of magnitude
predicted.

As previously stated, high sensitivity is very important for an imaging system to be useful in medical diagnosis.
The object being imaged must, of course, be insonified. If
the object is a living human organ, the power level of insonification must be low enough to avoid damaging the organ.
Since the piezoelectric systems offer the possibility of
producing effective images with the least tissue exposure to
ultrasound, these systems have a substantial advantage over
the laser systems. However, in applications other than
medical diagnosis, where operation at higher power is permissible, the laser systems may well be quite competitive,
particularly if phase retention is important.

Let us consider further the question of acoustic
orthographic imaging in real time as applied to medical
diagnosis. This is indeed a question of great interest.
The feasibility of using such imaging as a diagnostic tool
has already been demonstrated by Green et al [4]. The
necessity for developing practical real-time ultrasonic
imaging of orthographic character as an integral part of
diagnostic radiology has also been enunciated [5]. In
addition, the ability to obtain in real time good orthographic images of various parts of the human body without
using high acoustic intensity has been demonstrated by the
sensitive ultrasonic camera of Green and colleagues at
Stanford Research Institute [6]. As might be expected,
the system employs piezoelectric readout. Fig. 4 displays
one image from the SRI system showing part of the lower
leg. The bone and various blood vessels are seen in striking
detail. The center operating frequency for this image was
2 MHz, the instantaneous frequency being swept over a total
range of about 1 MHz (that is, from 1.5 to 2.5 MHz) to insure
incoherence and thereby eliminate speckle and ringing.

The SRI system is capable of imaging many internal
organs of the body trunk, including the colon, the kidneys,
the stomach and the heart. In the abdominal region, for
example, the spine, the lower ribs and the costal cartilage
can be viewed without difficulty. Muscles, tendons and
vascular structure are easy to identify. The fact that the
viewing is in real time and in true perspective (that is,
orthographic rather than cross-sectional) means that spatial
relationships between the various anatomical structures are
readily determined.

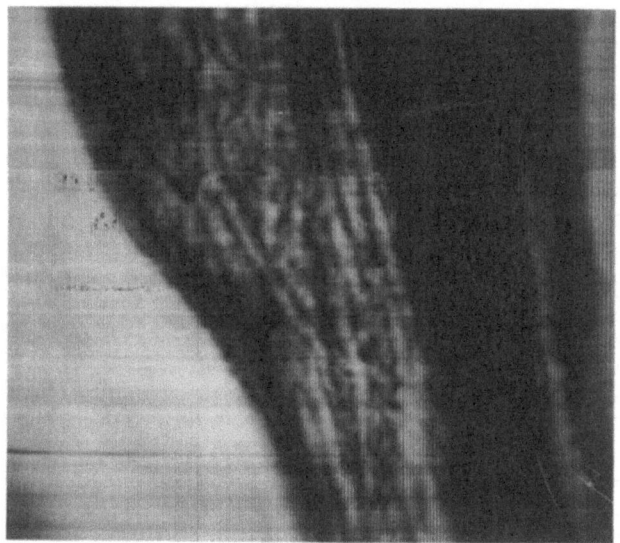

Fig. 4 Acoustic image of part of the lower leg obtained
by the ultrasonic camera of Stanford Research
Institute.

This system, with its piezoelectric readout, is aston-
ishingly sensitive. Employing a maximum peak pulse-power
density of only 18 mW/cm^2, the instrument will give a good
transmission image of the in vivo kidney of full-grown man.
The highest insonification intensity obviously occurs at
the surface of the skin facing the insonifying transducers.
As the waves penetrate the body, they are attenuated at a
rate of about 3 dB per cm. Because the system is pulsed
with a duty cycle of 1:60, the average power density even
at the entering skin surface is only 0.3 mW/cm^2. The
receiver, in this case an array of piezoelectric transducers,
has a threshold sensitivity of the order of 10^{-11} W/cm^2.
This means that the viewing of bodily organs can take place
through a considerable depth of tissue even though the
insonification intensity is well below that employed in
current diagnostic practice.

None of the laser-beams systems can approach the
sensitivity achieved by this piezoelectric system to within
two orders of magnitude. In addition, the SRI acoustic
camera, as can be seen from Fig. 4, is relatively free from
ringing and speckle. This fact supports the conclusions
previously reached, strongly indicating that a valid solution
to the problem of how to eliminate these unwanted interference products from the final image is to use non-coherent
sound.

III.   REFERENCES

[1] J. Landry, H. Keyani and G. Wade, "Bragg-diffraction
     imaging: a potential technique for medical diagnosis
     and material inspection," in Acoustical Holography,
     Vol. 4, G. Wade, Ed. New York: Plenum, 1972, pp. 127-
     146.

[2] A. Korpel, R. L. Whitman and M. Ahmed, "Elimination
     of spurious detail in acoustic images," in Acoustical
     Holography, Vol. 5, P. S. Green, Ed., New York: Plenum,
     1974, pp. 373-390.

[3] K. Wang and G. Wade, "Threshold contrast for three
     real-time acoustic imaging systems," in Acoustical
     Holography, Vol. 5, P. S. Green, Ed. New York: Plenum
     1974, pp. 239-247.

[4] P. S. Green, L. F. Schaefer and A. Macovski,
     "Considerations for diagnostic ultrasonic imaging,"
     in <u>Acoustical Holography</u>, Vol. 4, G. Wade, Ed.
     New York: Plenum, 1972, pp. 97-111.

[5] R. Anderson, "Potential medical applications for
     ultrasonic holography," in <u>Acoustical Holography</u>,
     Vol. 5, P. S. Green, Ed. New York: Plenum, 1974,
     pp. 239-247.

[6] P. S. Green, L. F. Schaefer, E. D. Jones and
     J. R. Suarez, "A new, high-performance ultrasonic
     camera," in <u>Acoustical Holography</u>, Vol. 5, P. S. Green,
     Ed. New York: Plenum, 1974, pp. 493-503.

Chapter 5

SCANNED ACOUSTIC HOLOGRAPHY

Byron B. Brenden

Holosonics, Inc.

2950 George Washington Way, Richland, WA 99352

5.1 INTRODUCTION

The initial and perhaps primary appeal of acoustical holography as a technique for imaging is that it circumvents the problem of the design and use of lenses. Relative to optical holography, this leaves the problem of replacing the photographic plates used as a detector with detectors more suitable for sensing and recording an acoustical hologram. Several suitable detectors have been devised[1-6] but perhaps the most obvious method for recording the acoustical hologram is the method whereby a point-like receiver is scanned over a plane in the zone of interference between two beams of ultrasound, one of which has interacted with the object. The second beam of ultrasound must be coherent with the source of ultrasound used to insonify the object. It must also have a spherical or plane wavefront since this facilitates duplicating the wavefront when an optical source is used to illuminate the hologram and form an image.

The basic situation is illustrated in Fig. 5.1. The elements used in this method are:

1)   A source to insonify the object,
2)   the object,
3)   a reference source, and
4)   a detector.

The detector, in this instance a focused piezoelectric transducer, is used to sample the interference pattern formed by the coherent addition of wave energy scattered

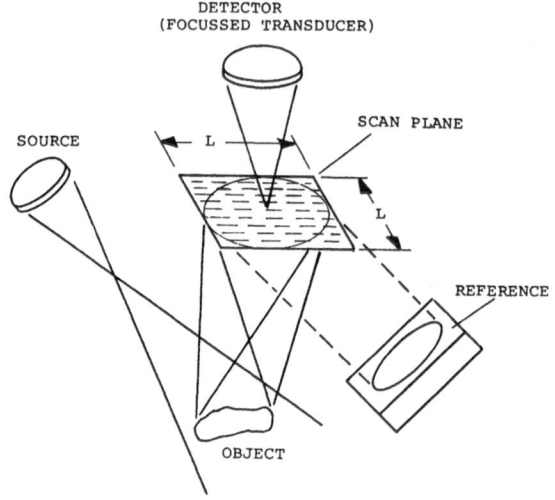

*Fig.5.1. The elements of a scanned acoustical holography
situation.*

from the object and wave energy in the reference beam. The
coupling media through which the sound is transmitted is
either a solid or a liquid. Practical applications for systems employing air or other gases as a coupling media are
exceedingly rare.

5.2 IMPROVED CONFIGURATIONS

The first improvement upon the scheme shown in Fig.
5.1 is to simulate the acoustic reference beam with an electronic signal. This is most readily done for an on-axis
plane wave. All that is required is to take a portion of
the signal used to drive the source and mix it with the signal derived from the detector. It is not much more difficult to simulate an off-axis plane wave and this procedure
has the advantage of allowing the object to be located directly below the scan plane. The object must always lie
outside the volume occupied by the reference beam whether
the reference beam be real or simulated. If it does not,
the image obtained upon illuminating the hologram will be
obscured by the light simulating the reference beam.

A second and less obvious step in simplifying the
scanned hologram situation is to eliminate the separate

source transducer and use the focussed piezoelectric transducer as both a source and receiver. This can be accomplished by driving the transducer with several cycles of
electrical energy at the desired frequency of operation and
then using the transducer as detector to sense all energy
reflected from the object. This technique is known as
simultaneous source-receiver scanning and, as shall be
shown, it provides a factor of two gain in resolution over
simple receiver scanning of the same size scan plane.

Having detected the signals corresponding to the interference pattern existing in the scan plane it is necessary
to use them to modulate the intensity of a point source of
light. The point source of light is used as a "pencil" to
draw the hologram in the face of a storage oscilloscope or
in the focal plane of a camera and must move in precisely
the same manner as the transducer. The end product is a
photographic transparency which is subsequently used in
forming an image.

5.3 BLOCK DIAGRAMS

The block diagram of the circuitry used to drive the
transducer and process the detected signal is shown in
Fig. 5.2. The signal from an oscillator is divided, part

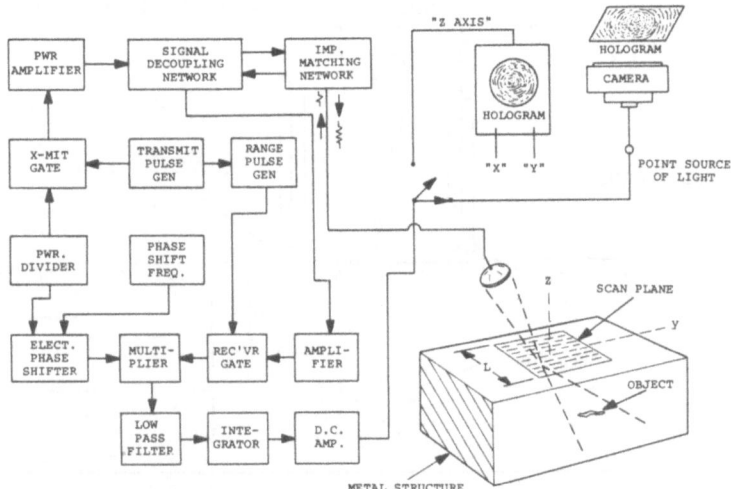

*Fig. 5.2. Electronic block diagram of a scanned acoustical
holography system.*

of it going to a transmit gate and part going to a phase
shifter. The first signal is used to drive the transducer.
The second provides a reference signal. A wave train of
approximately ten cycles is passed by the transmit gate and
after amplification is applied to the transducer to form a
transmit pulse.

The transducer, as illustrated in Fig. 5.2, is focussed
upon the surface of a thick plate of metal, such as the wall
of a reactor pressure vessel. The scan plane therefore lies
at the surface of the metal and the object is a flaw, a
void, crack or inclusion, lying deep within the metal structure. Energy reflected from this flaw passes back through
the surface and through the coupling medium between the surface and the transducer. The coupling medium is usually
water held in a small cylinder as illustrated in Fig. 5.3.
Fig. 5.3 also illustrates typical operating dimensions.
Transmit and receive gates can be set to span the flawed
volume while yet eliminating front and back surface reflections.

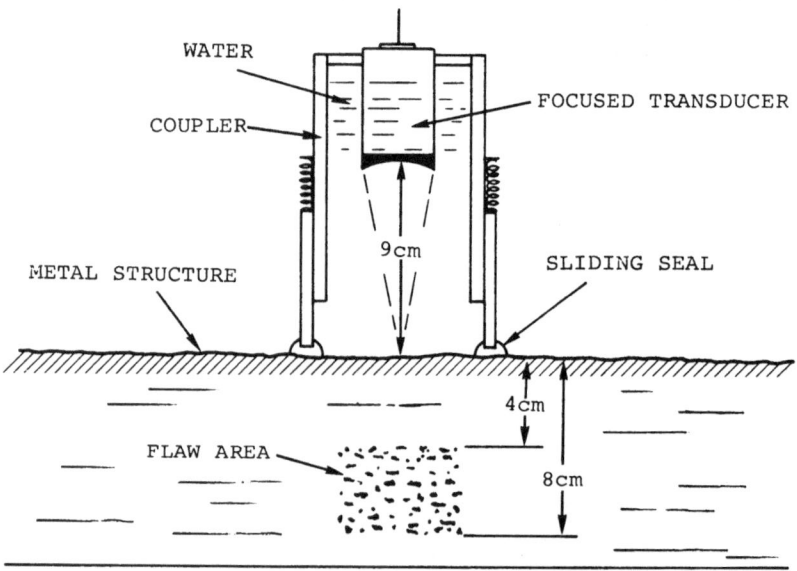

*Fig. 5.3. Method of coupling the scanning transducer to a
large solid surface.*

5.4 PRELIMINARY ANALYSIS

Returning again to Fig. 5.2, the signal derived from the reflected energy is amplified and filtered through a receive gate. Except for its time gated aspects it may now be characterized by the equation

$$V_1 = A \exp i \ (\Omega t + \phi_1(x,y)) \qquad (5.1)$$

where Ω is the angular frequency of the oscillator and $\phi_1(x,y)$ describes the phase variations characteristic of the object.

The reference signal may be represented by the equation

$$V_2 = B \exp i \ (\Omega t + \phi_2(x,y)) \qquad (5.2)$$

where the phase term $\phi_2(x,y)$ corresponds to that of a plane wave incident upon the scan plane at an angle Θ as shown in Fig. 5.4. A typical representation of ϕ_2 when the plane of indidence is the y-z plane would be

$$\phi_2(x,y) = \eta_2 y + \zeta_2 z \qquad (5.3)$$

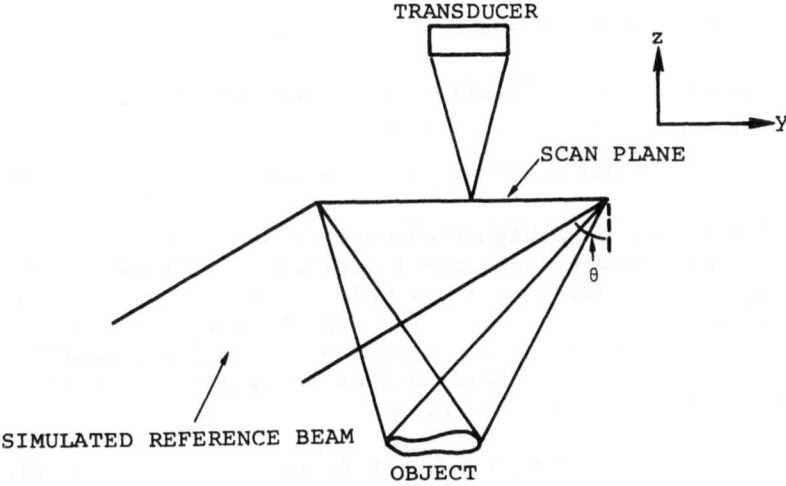

Fig.5.4. Schematic illustrating the geometrical equivalent of an electronically simulated reference beam.

where

$$\eta_2 = (2\pi/\Lambda)\sin\theta \qquad\qquad (5.4)$$

and

$$\zeta_2 = (2\pi/\Lambda)\cos\theta \qquad\qquad (5.5)$$

The two signals V_1 and V_2 could be added and squared as is done in optical holography by the photographic plate to produce the hologram characterized by the description

$$H = |V_1 + V_2|^2 = |V_1|^2 + |V_2|^2 + V_1 V_2^* + V_1^* V_2 \quad (5.6)$$

in which term

$$V_1 V_2^* = AB \exp i\ (\phi_1 - \phi_2) \qquad\qquad (5.7)$$

carries the information to form a true image.

In an alternate and preferred method, the two signals V_1 and V_2 can be multiplied and filtered. Taking the real parts of V_1 and V_2 as the inputs to the multiplier we have $AB\cos(\Omega t + \phi_1)\cos(\Omega t + \phi_2)$ as the output. But according to the trigometric identity

$$\cos x + \cos y = 2\cos(x + y)\cos(x - y) \qquad\qquad (5.8)$$

the output of the multiplier may be expressed:
$$AB\cos(\Omega t + \phi_1)\cos(\Omega t + \phi_2)$$
$$= AB[\cos(2\Omega t + \phi_1 + \phi_2) + \cos(\phi_1 - \phi_2)] \qquad (5.9)$$

The first term is a high frequency component of the signal which is removed by low pass filtering leaving only the desired signal $AB\cos(\phi_1 - \phi_2)$ which is the same as given in Eq. (5.7). As may be seen from Fig. 5.2, this is the signal used to modulate the brightness of the lamp, except that a DC bias is provided so that the signal again takes on the form of Eq. (5.6), namely:

$$H = H_o + V_1 V_2^* + V_1^* V_2 \qquad\qquad (5.10)$$

The film which is exposed to the light from the lamp may be processed to have a transmissivity which is linearly

related to the intensity of the lamp. When the film is
properly illuminated with coherent light characterized by
the function $V_a(\omega t)$, the transmitted light, V_b, is characterized by the product of the hologram transmissivity, H,
and the function V_a, i.e.,

$$V_b = V_a H \qquad (5.11)$$

That portion of the light that forms the true image is
characterized by the equation

$$V_t = V_a V_1 V_2^* \qquad (5.12)$$

A conjugate image is also formed. Light contributing to the
conjugate image is characterized by the equation

$$V_c = V_a V_1^* V_2 \qquad (5.13)$$

Figure 5.5 illustrates the manner in which an image
is formed. A beam of coherent light from a He-Ne laser is
expanded by a microscope objective lens to fill the aperture
of a 50mm diameter, 620mm focal length telescope objective
lens. The hologram is placed in a liquid gate near this

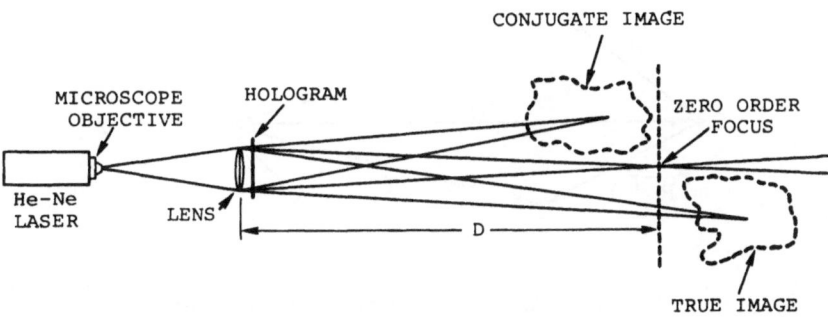

Fig. 5.5 Elements of the imaging system.

lens. The distance between the two lenses is adjusted so
that the zero order (undiffracted light) is focussed a distance D of approximately 10 meters from the hologram. Several diffracted orders are observed. The true image is
formed beyond the zero order focus in one of the two first
order diffracted beams.

5.5 SOURCE-RECEIVER RECIPROCITY

The recording of an acoustical hologram using a scanning receiver is a logical adaptation from optical holography wherein the photographic plate is replaced, or at
least displaced to a secondary role, by the piezoelectric
receiver. A somewhat less obvious modification of the
process is obtained when the roles of the source and the
receiver are interchanged [7,8]. To illustrate the rationale
behind source scanning refer to Fig. 5.6 in which a point
source S is located a vector distance $\bar{r}_s$, an object O is
located a vector distance $\bar{r}_0$ and a point receiver is located a vector distance $\bar{r}_r$ from the origin of an arbitrarily
located coordinate axis. The receiver generates a signal
V_0 which is mixed with a coherent reference signal V_r
given by

$$V_r = A \cos(\Omega t + \phi_r) \qquad (5.14)$$

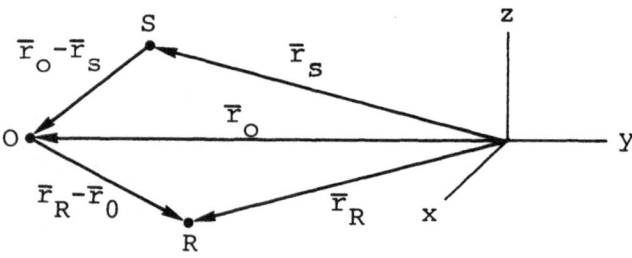

Fig. 5.6. The geometry of source-receiver reciprocity.

The signal V_o is made up of the sum of all the wave amplitudes reflected from different parts of the object, i.e.,

$$V_o = \int A(\bar{r}_o)\cos[\Omega t + \bar{k}_{so} \cdot (\bar{r}_o - \bar{r}_s) \\ + \bar{k}_{or} \cdot (\bar{r}_r - \bar{r}_o + \phi_o]d\bar{r}_o \qquad (5.15)$$

where $\bar{k}_{so}$ and $\bar{k}_{or}$ are wave propagation vectors having a magnitude of $2\pi/\bar{\Lambda}$ and $A(\bar{r}_o)$ is the amplitude of the acoustic wave scattered from the object at $\bar{r}_o$ to the receiver.

If the source and receiver positions are exchanged so that the source is at $\bar{r}_r$ and the receiver at $\bar{r}_s$, the new receiver signal V_o' is

$$V_o' = \int A(\bar{r}_o)\cos[\Omega t + \bar{k}_{ro} \cdot (\bar{r}_o - \bar{r}_r) \\ + \bar{k}_{os} \cdot (\bar{r}_s - \bar{r}_r) + \phi_o]d\bar{r}_o \qquad (5.16)$$

However,

$$\bar{k}_{ro} = -\bar{k}_{or} \text{ and } \bar{k}_{os} = -\bar{k}_{so}, \text{ so} \\ V_o' = V_o \qquad (5.17)$$

The signal V_r is not affected by this interchange. Thus, the resulting hologram is invariant to an exchange of the positions of the source and the receiver. It follows that holograms may be obtained by scanning the source instead of the receiver. The point source of light is still modulated by the low frequency component of the product of the reference and receiver signals but the motion of the light source now duplicates that of the source, rather than that of the receiver.

5.6 THE GEOMETRY OF IMAGING

Among the first questions to be answered with regard to acoustical holography is, "At what distance from the hologram do the images lie?" "Given an object distance r_1, a reference source distance r_2, and a reconstruction source distance r_a, as illustrated in Fig. 5.8, at what distance r_b may I expect to find the image?" The basic analytical

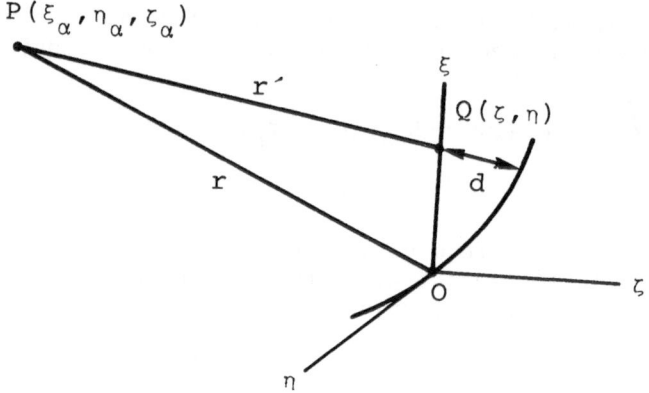

Fig.5.7.   Geometry of the scan situation.

study of this question together with a treatment of third order aberrations has been given by Meier[9] and Champaigne[10]. These analyses treat the object as if it were a single point. A more general treatment, in which the object is considerd to be a complex structure is given by Hildebrand and Brenden[11]. For the purpose of this chapter the use of a single point object will suffice.

Consider the general wave diagram of Fig. 5.7. The point P may represent one of several entities, namely, the reference source, a point in the object, the insonifying source, the reconstruction source or a point in the image. Wave energy emanating from the point P may be described by the function

$$V = A \exp i(\Omega t - \bar{k} \cdot \bar{r})$$

$$= A \exp i(\Omega t - \phi) \qquad (5.18)$$

$$= U \exp i \Omega t$$

where

$$U = \exp(-i\phi) = A \exp [-i K (r' - r)] \qquad (5.19)$$

In Eq. (1.18) $\bar{k}$ is the wave propagation vector and $\bar{r}$ is
the vector from the origin of the coordinate system to the
point P. In Eq. (5.19) a scalar product is intended and
the distance r' - r is equivalent to the distance d in Fig.
5.7.

The time dependent term exp(iΩt) cancels out of all
equations and, therefore, is omitted in the discussion
from this point on. The expression equivalent to Eq. (5.12)
now becomes

$$U_T = A_a A_1 A_2 \exp [-i (\phi_1 - \phi_2 + \phi_a)] \tag{5.20}$$

All the geometrical relationships are derived from the
phase term. Our first problem is to express the phase in
terms of the coordinates ξ, η and ζ. To do this, we note
that

$$r'_\alpha = [(\xi - \xi_\alpha)^2 + (\eta - \eta_\alpha)^2 + \zeta_\alpha^2]^{\frac{1}{2}} \tag{5.21}$$

and that

$$r_\alpha = [\xi_\alpha^2 + \eta_\alpha^2 + \zeta_\alpha^2]^{\frac{1}{2}} \tag{5.22}$$

Equation (5.21) can be written in form

$$\begin{aligned} r'_\alpha &= [\xi^2 - 2\xi\xi_\alpha + \xi_\alpha^2 + \eta^2 - 2\eta\eta_\alpha + \eta_\alpha^2 + \zeta_\alpha^2]^{\frac{1}{2}} \\ &= [r_\alpha^2 + \xi^2 + \eta^2 - 2\xi\xi_\alpha - 2\eta\eta_\alpha]^{\frac{1}{2}} \end{aligned} \tag{5.23}$$

Using the bionominal expansion (or Taylor's series expansion) we recall that

$$(a + b)^n = a^n + a^{n-1} b + \tfrac{1}{2}n(n-1)a^{n-2} b^2 \ldots \tag{5.24}$$

If we set $n = \frac{1}{2}$, $a = r^2$ and $b = \xi^2 + \eta^2 - 2\xi\xi_\alpha - 2\eta\eta_\alpha$ we
find that

$$r'_\alpha - r_\alpha = (\xi^2 + \eta^2)/2r_\alpha - (\xi\xi_\alpha + \eta\eta_\alpha)/r_\alpha + 0(\xi^4/r^3) \tag{5.25}$$

where $0(\xi^4/r^3)$ indicates terms of the magnitude ξ^4/r^3 or
less. These terms will be dropped except for the derivation of third order aberration coefficients since it is

assumed that we limit our analysis to cases for which
$\xi<<r$ and $\eta<<r$.

Consider Fig. 5.8 in which the point $OB(\xi_1,\eta_1,\zeta_1)$ is a
point in the object and the point $KS(\xi_2,\eta_2,\zeta_2)$ is the
source of the reference signal. Electronically simulated
reference beams are normally the equivalent of plane waves
in which case RS is a point source at infinity, i.e., r_2 is
infinite.

The point $Q(\xi,\eta)$ represents a point receiver lying in
the ξ,η plane. Normally Q would be the scanned element but,
because of source-receiver reciprocity, the receiver could
be held in a fixed position while the source point $OS(\xi_s,
\eta_s,\zeta_s)$ is scanned. Another mode of operation, as will be
shown in the discussion to follow, is to scan both the
source, OS, and the receiver, Q, together. This mode is
called simultaneous source-receiver scanning. In order to
describe scanning arrangements in which both the receiver
and the transmitter scan simultaneously we must choose func-
tions which specify ξ_s,η_s,ζ_s, in terms of ξ, η and ζ. One

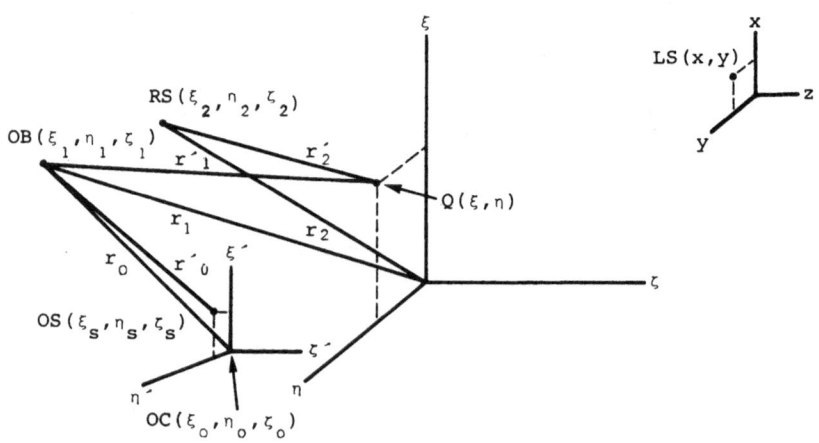

*Fig.5.8. Parameters relating to the formation
of a hologram.*

of the most useful specifications is provided by the equations

$$\xi_s = p\xi + \xi_o \; , \quad \eta_s = p\eta + \eta_o \quad \text{and} \quad \zeta_s = 0 \qquad (5.26)$$

in which p is a constant.

The point source of light LS(x,y), moves in its own scan plane, the x,y plane. Its motion is a magnified version of the motion of the scanning element in the ξ,η plane. The magnification, m, is usually less than unity. If we take m to be the product of any mechanical, electrical and optical magnifications so that the x,y plane is really the film plane in which the hologram is recorded then

$$\xi = x/m \quad \text{and} \quad \eta = y/m \qquad (5.27)$$

We should now correct Eq. (5.20) to read

$$U_T = A \exp \; [-i \; (\phi_s + \phi_1 - \phi_2 + \phi_a)] \qquad (5.28)$$

in which the amplitude factors have been lumped into A and in which the phase, ϕ_s, of the source is taken into account. Similarly in accordance with Eq. (5.13)

$$U_c = A \exp \; [i \; (\phi_s + \phi_1 - \phi_2 - \phi_a)] \qquad (5.29)$$

The ϕ_s term differs somewhat in form from that given by Eq. (5.25). Note that

$$r'_o = [(\xi_s - \xi_1)^2 + (\eta_s - \eta_1)^2 + (\zeta_s - \zeta_1)^2]^{\frac{1}{2}} \qquad (5.30)$$

and

$$r_o = [(\xi_o - \xi_1)^2 + (\eta_o - \eta_1)^2 + (\zeta_o - \zeta_1)^2]^{\frac{1}{2}} \qquad (5.31)$$

so that

$$r'_o - r_o = p^2(\xi^2 + \eta^2)/2r_o - p \; \xi(\xi_1 - \xi_o)/r_o \qquad (5.32)$$
$$- p \; \eta(\eta_1 - \eta_o)/r_o + 0(\xi^4/r^3)$$

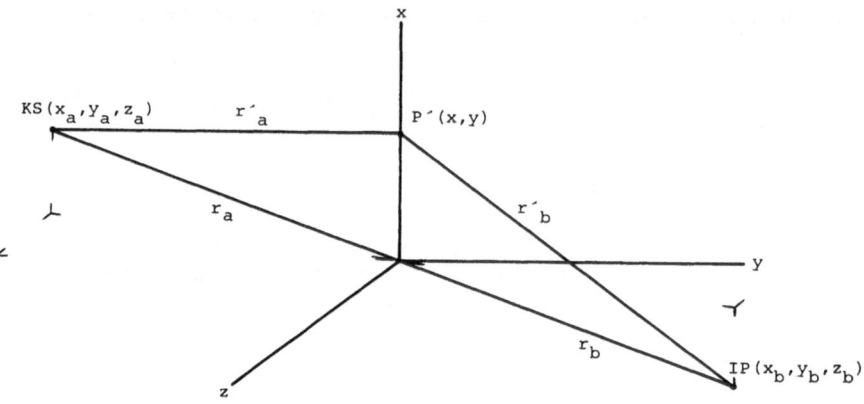

Fig. 5.9. Parameters relating to the formation of an image using a hologram.

If we substitute from Eqs. (5.26) and (5.27) into Eq. (5.32) we see that

$$r'_o - r_o = \frac{p^2(x^2 + y^2)}{2m^2 r_o} - \frac{px(\xi_1 - \xi_o)}{mr_o} - \frac{py(\eta_1 - \eta_o)}{mr_o} \qquad (5.33)$$

Just as Fig. (5.8) gives the geometry for forming a hologram so Fig. (5.9) gives the geometry for forming an image using a hologram. The phase of the reconstructed wave U_T or U_c, which forms the image point IP is given by

$$\phi_{T/c} = k(r'_\alpha - r_\alpha) \pm K(r'_o - r_o + r'_1 - r_1 - r'_2 + r_2) \qquad (5.34)$$

where

$$K = 2\pi/\Lambda \quad \text{and} \quad k = 2\pi/\lambda \qquad (5.35)$$

where Λ is the wavelength of the acoustical wave used to form the hologram and where λ is the wavelength of the light used to form an image. Making the appropriate substitutions into Eq. (5.34) we obtain

$$\phi_{T/c} = k[(x^2 + y^2)/2r_a - xx_a/r_a - yy_a/r_a]$$

$$\pm K[p^2(x^2 + y^2)/2m^2r_o - px(\xi-\xi_o)/mr_o$$

$$-py(\eta-\eta_o)/mr_o + (x^2 + y^2)/2m^2r_1 \qquad (5.36)$$

$$-x\xi_1/mr_1 - y\eta_1/mr_1 - (x^2 + y^2)/2m^2r_2$$

$$+x\xi_2/mr_2 + y\eta_2/mr_2]$$

A perfect hologram would diffract light from the reconstruction source KS in such a way that light converging on the image point IP would have a perfectly spherical wavefront describable in terms of phase by

$$\phi_b = k(r'_b - r_b) \qquad (5.37)$$

or

$$\phi_b = k[(x^2 + y^2)/2r_b - xx_b/r_b - yy_b/r_b] + O(x^4/r^3) \qquad (5.38)$$

The actual wavefront is given by Eq. (5.38). Imaging occurs under the condition that

$$\phi_{T/c} \approx - \phi_b \qquad (5.39)$$

Any difference $\phi_b + \phi_{T/c}$ represents an imaging defect or aberration. The minus sign is used with ϕ_b in Eq. (5.39) because we wish to establish the sign convention used in optics that the distance, r_a, is positive if energy flows from the source KS to the hologram and the distance r_b is positive if the energy flows from the hologram to the image point IP.

Comparison of Eqs. (5.36) and (5.38) shows that the conditions for forming the true image are given by

$$-1/r_b = (\lambda/m^2\Lambda)(p^2/r_o + 1/r_2) + 1/r_a \qquad (5.40)$$

$$-x_b/r_b = (\lambda/m\Lambda)\,[p(\xi_1 - \xi_o)/r_o + \xi_1/r_1 - \xi_2/r_2] + \xi_a/r_a \quad (5.41)$$

and

$$-y_b/r_b = (\lambda/m\Lambda)\,[p(\eta_1 - \eta_o)/r_o + \eta_1/r_1 - \eta_2/r_2] + \eta_a/r_a \quad (5.42)$$

5.7 MAGNIFICATION

The preceding equations allow us to calculate radial and lateral magnifications. For the radial magnification

$$(5.43)$$

$$M_R = dx_b/dr_1 = (\lambda/m^2\Lambda)(r_b/r_1)^2[1 + p^2(r_1/r_o)^2 dr_o/dr_1]$$

A particular case of interest is one in which the source and the receiver are coincident in which case $r_o = r_1$, $dr_o/dr_1 = 1$ and $p = 1$.

These conditions describe a simultaneous source-re-ceiver scan system. Note that the radial magnification is a factor of two greater for such a system than for one in which only the receiver is moved, i.e., for which $p = 0$.

The lateral magnification is given by

$$M_L = dx_b/d\xi_1 = (\lambda/m\Lambda)(r_b/r_1)[1 + p(r_1/r_o)] \quad (5.44)$$

where it is assumed in taking the derivative that $x_b \ll r_b$ and $\xi_1 \ll r_1$. Again, the lateral magnification is twice as great for the simultaneous source-receiver scan case ($p = 1$) as for the receiver scan case ($p = 0$). When $p = 1$

$$M_R/M_L = (\Lambda/2\lambda)M_L \quad (5.45)$$

and since the ratio of wavelengths is in general very large the radial dimensions of the image are greatly stretched relative to the lateral dimensions. The same situation holds in ordinary optical microscopy.

5.8 RESOLUTION

The resolving power of an imaging system is defined in terms of the least separation, $\Delta\xi_1$, between two points in the object which can be detected in the image. In the case of holograms, we can define resolving power as the distance, $\Delta\xi_1$, which an object point may be moved before the phase of the finest fringe in the hologram is reversed. The phase of the energy forming the hologram is given in Eq. (5.38). The incremental change in phase caused by an incremental change in the position of the object point OB (Fig. 5.8) is

$$\Delta\phi = (\delta\phi/\delta\xi_1)\Delta\xi_1 \simeq (Kx/m)(p/r_o + 1/r_1)\Delta\xi_1 \qquad (5.46)$$

The phase of the finest fringe is reversed when $\Delta\phi = \pi$. Thus the least detectable movement, $\Delta\xi_1$, of the object point is

$$\Delta\xi_1 = \Lambda/L(p/r_o + 1/r_1) \qquad (5.47)$$

where the width of the scan plane L (See Fig. 5.1) has been substituted for the equivalent expression $2x_{max}/m$. Eq. (5.47) shows that the resolution is twice as good for the simultaneous source-receiver scan case ($p = 1$ and $r_o = r_1$) as for the receiver scan case ($p = 0$).

5.9 SIMULTANEOUS SOURCE-RECEIVER SCAN

As previously discussed the simultaneous source-receiver scan technique offers the advantages of better resolution by a factor of 2 for a given size scan plane. In addition, the technique has the practical advantages of utilizing only a single transducer and of insonifying the object from a variety of angles. Equation (5.40) can be put in the form suitable for simultaneous source-receiver scanning by setting $p = 1$, $r_o = r_1$ and $\xi_o = \eta_o = 0$. Furthermore only plane wave reference beams will be considered so that $r_b \rightarrow \infty$. Then the equations for the true image become

$$-1/r_b = 2\lambda/m^2 r_1\Lambda + 1/r_a \qquad (5.48)$$

If in addition we express x/r or ξ/r as $\cos\alpha$ as illustrated in Fig. 5.10, Eq. (5.41) becomes

$$-\cos\alpha_b = (\lambda/m\Lambda)(2\cos\alpha_1 - \cos\alpha_2) + \cos\alpha_a \qquad (5.49)$$

With β defined as shown in Fig. 5.10, Eq. (5.42) becomes

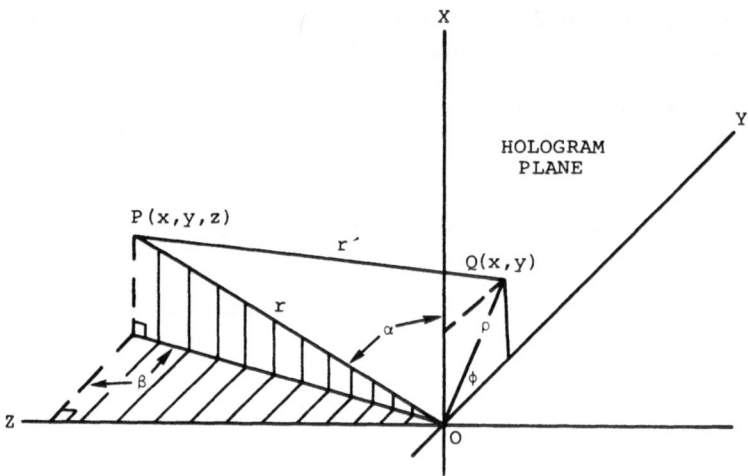

Fig.5.10. Parameters used in defining wave front aberrations.

$$-\sin \alpha_b \sin \beta_b = (\lambda/m\Lambda)(2 \sin \alpha_1 \sin \beta_1 \qquad (5.50)$$
$$- \sin \alpha_2 \sin \beta_2) + \sin \alpha_a \sin \beta_a$$

5.10 ABERRATIONS

The previous equations have all been derived on the basis of neglecting terms of magnitude x^4/r^3 or smaller. When terms of order x^4/r^3 are included in the anlaysis an expression for the wave aberration, Δ, can be derived. This expression is

$$\Delta = \frac{1}{\lambda} \left[-\frac{1}{8} \rho^4 S + \frac{1}{2} \rho^3 (C_x \cos \phi + C_y \sin \phi) \qquad (5.51)\right.$$
$$\left. -\frac{1}{2} \rho^2 (A_x \cos^2 \phi + A_y \sin^2 \phi + 2A_{xy} \sin \phi \cos \phi)\right]$$

in which

$$\rho^2 = x^2 + y^2 \qquad (5.52)$$

The quantities S, C, A and $A_{x,y}$ are called the coefficients for spherical aberration, coma, astigmatism and curvature of field respectively. These aberration coefficients may be calculated from[11]

$$S = (\lambda/m^4\Lambda)(2/r_1^3 - 1/r_2^3) + 1/r_a^3 + 1/r_b^3 \qquad (5.53)$$

$$C_x = (\lambda/m^3\Lambda)[(2 \cos \alpha_1/r_1^2) - (\cos \alpha_2/r_2^2)]$$
$$+ (\cos \alpha_a/r_a^2) + (\cos \alpha_b/r_b^2) \qquad (5.54)$$

$$C_y = (\lambda/m^3\Lambda)](2 \sin \alpha_1 \cos \beta_1/r_1$$
$$- (\sin \alpha_2 \cos \beta_2/r_2^2)] +$$
$$+ (\sin \alpha_a \cos \alpha_a/r_a^2)$$
$$+ (\sin \alpha_b \cos \beta_b/r_b^2) \qquad (5.55)$$

$$A_x = (\lambda/m^2\Lambda)[(2 \cos^2 \alpha_1/r_1) - (\cos^2 \alpha_2/r_2)]$$
$$+ (\cos^2 \alpha_a/r_a) + (\cos^2 \alpha_b/r_b) \qquad (5.56)$$

$$A_y = (\lambda/m^2\Lambda)[(2 \sin^2\alpha_a \cos^2\beta_1/r_1)$$
$$- (\sin^2\alpha_2 \cos^2\beta_2/r_2)]$$
$$+ (\sin^2 \alpha_a \cos^2\beta_a/r_a)$$
$$+ (\sin^2\alpha_b \cos^2\beta_b/r_b) \qquad (5.57)$$

$$A_{xy} = (\lambda/m^2\Lambda)[(2 \sin \alpha_1 \cos \alpha_1 \sin \beta_1/r_1)$$
$$- (\sin \alpha_2 \cos \alpha_2 \sin \beta_2/r_1)]$$
$$+ (\sin \alpha_2 \cos \alpha_a \sin \beta_a/r_a)$$
$$+ (\sin \alpha_b \cos \alpha_b \sin \beta_b/r_b \qquad (5.58)$$

REFERENCES

1.   R. K. Mueller and N. K. Sheriden, Appl. Phys.
     Lett. 9: 328-329 (1966)

2.   B. B. Brenden and R. B. Kidman, U. S. Patent No.
     3,683,679 (1972)

3.   J. D. Young and J. E. Wolfe, Appl. Phys. Lett.
     11:9 (1967)

4.   A. F. Metherell, Appl. Phys. Lett. 13:10 (1968)

5.   A. Korpel and P. Demares, J. Acoust. Soc. Am.
     45:4 (1969)

6.   R. B. Smith and B. B. Brenden, Ultrasonics
     7:125-126 (1967)

7.   A. F. Metherell and S. Spinak, Appl. Phys. Lett.
     13:22 (1968)

8.   V. I. Neeley, Phys. Lett. 28A(7):475-476 (1968)

9.   R. W. Meier, J. Opt. Soc. Am. 55(8):987 (1965)

10. E. B. Champagne, J. Opt. Soc. Am. 57(1):51 (1967)

11. B. P. Hildebrand and B. B. Brenden, An Introduction
     to Acoustical Holography, Plenum Press, New York
     (1972) Plenum/Rosetta edition Sec. 2.6

Chapter 6

LIQUID SURFACE HOLOGRAPHY

Byron B. Brenden

Holosonics, Inc.

2950 George Washington Way, Richland, WA 99352

6.1 INTRODUCTION

As an acoustical hologram detector, the liquid surface
has the distinct advantage that no scanning process is required. The interaction of two acoustical waves at a liquid
surface occurs uniformly over a period of approximately 100
microseconds and produces deformations on the liquid surface
that are capable of holographically reconstructing an optical wave equivalent to one of the two acoustical waves. The
process may be repeated at the rate of sufficient to produce
nearly 200 holograms and images per second. This is a
sufficiently high rate of hologram production to allow as
many as ten different frequencies to be used with the images
of each frequency occurring rapidly enough to exceed the
critical fusion frequency of the eye.

A schematic showing one possible layout of a practical
liquid surface imaging system is shown in Fig. 6.1. This
schematic drawing shows an object transducer generating a
beam of acoustical energy which interacts with an object
to be imaged. The object is brought into acoustical contact with the energy by means of a flexible water filled
bag. A similar bag is used on the opposite side of the object. The acoustic wave transmitted through the object is
altered in phase and amplitude and thereby carries information about the object which, in the system defined by Fig.
6.1, is conveyed to the imaging tank by a pair of acoustic
lenses. The upper lens can be set at one focal length
from the liquid surface in the imaging tank. The lower lens

can then be moved up and down to bring different planes
in the object into focus.

The interference pattern derived from mixing a plane
wave reference beam with the object beam at the surface of
the liquid in the imaging tank produces deformations on the
surface so that the surface becomes a complex phase grating.
Light, from a pulsed argon-ion laser, focused to a point and
collimated by a collimating lens, illuminates the liquid
surface with pulses of approximately 10 microseconds duration at some optimum time after the hologram is fully formed.
Light reflected from the surface is diffracted into several
orders. The undiffracted, zero order light is blocked at
the spatial filter. A video camera, focussed on the liquid
surface through the collimating lens, forms an image of a
selected plane within the object on the television monitor.

It is apparent from the above description that the system of Fig. 6.1 forms a focused image hologram which is
immediately and instantaneously used to form an optical
image. Because a hologram is formed, it is not necessary to
focus an acoustic image precisely in the plane of the liquid
surface. The acoustic image may exist on either side of

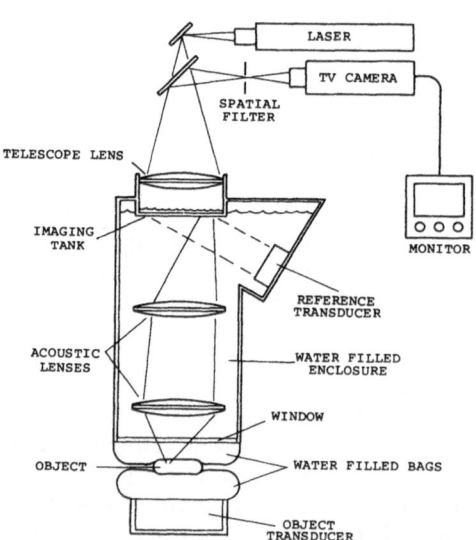

*Fig.6.1. One arrangement used in liquid surface acoustical
holography.*

the liquid surface with the video camera focused accordingly. However, it will be noted by application of Eq. (5.39)
that the distance, r_b, of the optical image from the hologram is given by

$$r_b = -(\Lambda/\lambda)r_1 \tag{6.1}$$

For ultrasound at 5MHz in water $\Lambda = 0.3 \times 10^{-3}$m and for an
argon-ion laser $\lambda \simeq 0.6 \times 10^{-6}$m so

$$r_b = -500r_1 \tag{6.2}$$

Thus, even for relatively small values of r_1, such as 1cm,
the optical image lies at nearly 10 focal lengths from the
collimater lens. The collimater lens then images the hologram image in a plane near the spatial filter. Note that
r_b, and therefore the location of the optical image, is dependent upon the acoustical wavelength, Λ. Only when $r_1 = 0$,
i.e., when the acoustical image lies in the plane of the
hologram do we have the same focus for all frequencies of
ultrasound. Focused image holography is therefore a convenient way to accommodate multifrequency operation. Nevertheless, our discussion of liquid surface holography will
begin with the more basic, lenseless system.

6.2 DESCRIPTION OF THE ACOUSTICAL FIELD

Acoustical energy, bounded in space and emitted in
wave trains several hundred wavelengths long, is used in
liquid surface systems for imaging. Waves bounded in space
and time can be treated as a linear superposition of unbounded plane waves. The analysis of the response of a
liquid surface detector to acoustical energy will begin on
the basis of unbounded waves.

Figure 6.2 depicts certain features of the imaging
situation. We shall consider two acoustic waves, one incident upon the surface at an angle Θ_1 and the other at an
angle Θ_2. These two waves may be characterized by the
equations

$$U_1 = P_1 \exp[-i(\eta_1 y - \zeta_1 z + \phi)] \tag{6.3}$$

and

$$U_2 = P_2 \exp[i(\eta_2 y + \zeta_2 z)] \tag{6.4}$$

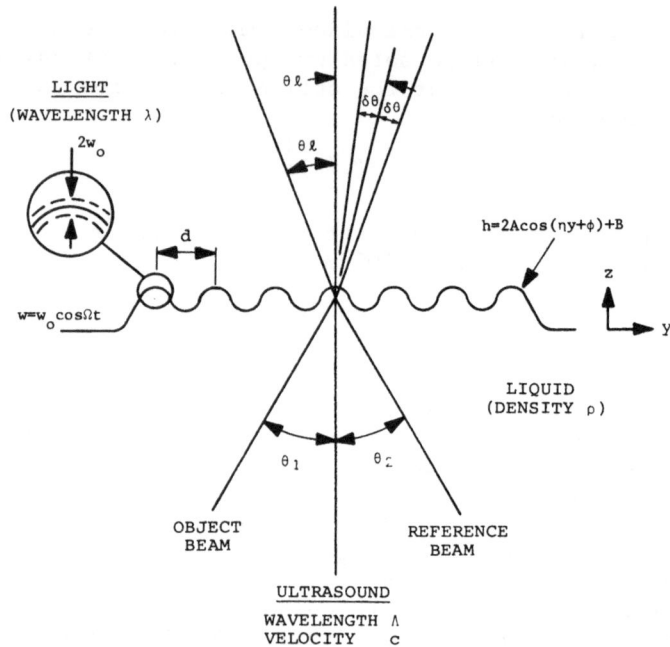

Fig.6.2. Parameters used in discussing liquid surface holography.

where p_1 and p_2 are the pressure amplitudes of the waves and where

$$\eta_i = (2\pi/\lambda) \sin \theta_i \qquad (6.5)$$

and

$$\zeta_i = (2\pi/\lambda) \cos \theta_i \qquad (6.6)$$

In the formulation given above both θ_1 and θ_2 are considered positive as shown in Fig. 6.2. U_1 may be considered to be one component of the acoustical beam that has passed through the object. The added phase term Θ characterizes the object. At the liquid surface where $z = 0$ the intensity distribution in the interference pattern is given by

$$I = |U_1 + U_2|^2/2\rho c \qquad (6.7)$$

where ρ is the density of the liquid and c is the velocity of sound in the liquid. Reflection of the sound at the liquid surface produces a radiation pressure given by

$$\Pi = 2I/c \qquad (6.8)$$

This pressure is opposed by pressures due to gravity and surface tension. That due to gravity may be calculated simply as the weight of a column of liquid of unit cross-sectional area and of height z, i.e.,

$$\Pi_g = -\rho g z \qquad (6.9)$$

The pressure, Π_t, due to surface tension may be expressed[1] as

$$\Pi_t = \gamma(\partial^2 z/\partial y^2) \qquad (6.10)$$

Thus, the local pressure balance equation becomes

$$\Pi = \rho g z - \gamma(\partial^2 z/\partial y^2) \qquad (6.11)$$

Referring again to Eq. (6.8) note that for

$$|(\zeta_1 - \zeta_2)z| \ll |(\eta_1 + \eta_2)y| \qquad (6.12)$$

$$|U_1 + U_2|^2 = P_1^2 + P_2^2 + 2P_1 P_2 \cos(\eta y + \phi) \qquad (6.13)$$

where

$$\eta = \eta_1 + \eta_2 \qquad (6.14)$$

Assuming a stationary solution for z in the form

$$z(y) = 2A\cos(\eta y + \phi) + B \qquad (6.15)$$

and applying Eq. (6.11) yields

$$A = P_1 P_2/\rho c^2(\rho g + \gamma \eta^2) \qquad (6.16)$$

and

$$B = (P_1^2 + P_2^2)/\rho^2 c^2 g \qquad (6.17)$$

In many particular cases $\rho g \ll \gamma \zeta \eta^2$ so that Eq. (6.16) is normally well approximated by

$$A = P_1 P_2/\gamma \rho c^2 \eta^2 \qquad (6.18)$$

In Fig. 6.2 the distance d may be calculated from

$$d = 2\pi/\eta \qquad (6.19)$$

6.3 INTERACTION OF LIGHT WITH THE LIQUID SURFACE

The acoustical field pattern in the object will be translated into an equivalent optical image if each plain wave component U_1 of the object beam is translated into an equivalent optical wave. The light incident upon the liquid surface may be characterized by the equation

$$U_a = D \exp [i(\eta_\ell y - \zeta_\ell z)] \qquad (6.20)$$

which describes a plane wave of amplitude D incident upon the liquid surface at an angle Θ_ℓ to the normal as illustrated in Fig. 6.2. The wavelength of the light is λ so

$$\eta_\ell = (2\pi/\lambda) \sin \theta_\ell \qquad (6.21)$$

and

$$\zeta_\ell = (2\pi/\lambda) \cos \theta_\ell \qquad (6.22)$$

To avoid confusing the coordinate z with the liquid surface elevation as given by Eq. (6.15) we define a function h such that

$$h = z(y) \qquad (6.23)$$

For sufficiently small values of Θ_ℓ (i.e., for light incident at near normal to the surface) the difference in path distance between a ray reflected from the quiescent surface $z = 0$ and the hologram surface is 2h. The reflected wave should therefore have the form

$$U_b = R D \exp [i(\eta_\ell y + \zeta_\ell (z - 2h))] \qquad (6.24)$$

where R is the amplitude reflection coefficient. Substituting from Eq. (6.15) into Eq. (6.24) and re-arranging terms yields the expression

$$U_b = R D \exp [2i\zeta_\ell B] \exp [4i\zeta_\ell A \cos (\eta y + \phi)]$$
$$\times \exp [i(\eta_\ell y + \zeta_\ell z)] \qquad (6.25)$$

There is an identity, namely

$$\exp (i \, \sigma \cos \alpha) = \sum_{n} i^{n} J_{n}(\sigma) \exp (-in\alpha) \qquad (6.26)$$

which is useful for facilitating the physical interpretation of Eq. (6.25). The summation includes integral values of n from $-\infty$ to $+ \infty$.

Let

$$\sigma = 4\zeta_{\ell} A \qquad (6.27)$$

and

$$\alpha = n_{\ell} y + \phi \qquad (6.28)$$

Using the identity, Eq. (6.25) becomes

$$U_{b} = R \, D \, \exp [i(\zeta_{\ell} z + 2B)] \sum_{n} i^{n} J_{n} (4\zeta_{\ell} A) \qquad (6.29)$$

$$x \, \exp [i((n_{\ell} - n\eta)y - n\phi)]$$

Each value of n corresponds to a diffracted order of light. When the argument, $4\zeta_{\ell} A$ is sufficiently small Eq. (6.29) can be approximated by

$$U_{b} = R \, D \, \exp [i\zeta_{\ell} (z + 2B)] \{ \exp i \, n_{\ell} y \qquad (6.30)$$

$$+ 2i\zeta_{\ell} A \, \exp [i((n_{\ell} - \eta)y - \phi)]$$

$$+ 2i\zeta_{\ell} A \, \exp [i((n_{\ell} + \eta)y + \phi)]$$

$$\equiv U_{o} + U_{T} + U_{c}$$

If

$$n_{\ell} = n_{2} \qquad (6.31)$$

then the optical wave, U_{T}, describes the image of the acoustical wave, U_{1}. Even if the condition of Eq. (6.31) is not met, good images are obtained, the only effect being a difference in magnification between the x and y dimensions.

Substituting for A from Eq. (6.18) we find that

$$U_{b} U_{b}^{*} = (2RD\zeta_{\ell} P_{2} / \gamma \rho c^{2} \eta^{2})^{2} P_{1}^{2} \qquad (6.32)$$

Using $\dfrac{p^2}{2\rho c}$ for the intensity of an acoustic beam Eq. (6.32)
may be rewritten

$$I_b = (16RD\zeta_\ell / \gamma c^2 n^4)\, I_1 I_2 \qquad (6.33)$$

The ratio I_b/I_1 is a measure of the sensitivity of the detector. Evidently, liquids of low surface tension, high reflectivity and low acoustic velocity are to be desired. Furthermore, the intensity of the reference beam should be maintained as high as possible.

6.4   THE DYNAMICS OF LIQUID SURFACE MOTION

The magnitudes of A and B given by Eqs. (6.16) and (6.17) are the steady state magnitudes. The pulsing of the waves U_1 and U_2 brings into play inertial forces which greatly alter the relative magnitudes of A and B.

It is characteristic of simple harmonic motion that the frequency, f, the mass, M, the amplitude, G, and the total energy, E, are related through the equation

$$f = (1/2\pi G)\,\sqrt{2E/M} \qquad (6.34)$$

Although the oscillations induced in a liquid surface by the repetitive wave trains used in imaging may not be truly simple harmonic in nature, Eq. (6.34) is useful in elucidating the dynamic characteristics of liquid surface response. There are three modes of vibration to be considered, one related to the amplitude, w_o, of the particle displacement, one related to the amplitude, B, which we shall call the bulge amplitude, and one related to the amplitude, A, of the liquid surface grating. The first mode of vibration is simple harmonic in nature and needs no further discussion.

For the bulge

$$E = \frac{1}{2}\,MgB \qquad (6.35)$$

and

$$G = B \qquad (6.36)$$

so that according to Eq. (6.34), the natural frequency of the bulge is

$$f_b = (1/2\pi) \sqrt{g/B} \qquad (6.37)$$

Equation (6.17) may be written in the form

$$B = 2(I_1 + I_2)/\rho cg \qquad (6.38)$$

Taking a value of 4×10^{-4} watts/cm^2 for the average value of the numerator and using $\rho = 1.79 g/cm^3$, $c = 0.73 \times 10^5$ cm/sec^2 and $g = 980 cm/sec^2$ we find that

$$B \approx 3 \times 10^{-5} cm \qquad (6.39)$$

and

$$f_b = 890 \text{ Hz} \qquad (6.40)$$

The natural frequency of the ripple pattern of amplitude A is determined in a similar manner. In this instance the restoring force for the oscillating system is surface tension. The total energy, E, or Eq. (6.34) is the work done against surface tension in forming the ripple pattern. Consider Fig. 6.3. Let s be the distance along the surface from point P1 to point P2 when the ripple pattern has reached its maximum amplitude 2A. The distance between the same two points is d before the surface is distorted. The work done per unit depth in the x-direction in distorting the surface is

$$W = (s-d)\gamma \qquad (6.41)$$

where γ is the surface tension. To calculate the work we must first calculate s.

An elemental path length along the ripple surface is given by

$$ds = [(dz)^2 + (dy)^2]^{1/2} \qquad (6.42)$$

thus

$$ds/dy = [1 + (dz/dy)^2]^{1/2} \approx 1 + \frac{1}{2}(dz/dy)^2 \qquad (6.43)$$

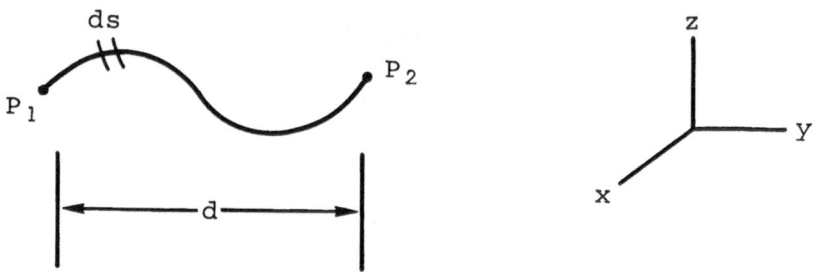

Fig.6.3. Parameters relating to path length along a rippled surface.

substituting from Eq. (6.15) for z and integrating we find that

$$s = d + 2\pi A^2 \eta \qquad (6.44)$$

Thus the work done per unit depth in the x-direction to deform the surface is

$$W = 2\pi A^2 \eta \gamma \qquad (6.45)$$

Equating the total energy E to W and substituting for η from Equation (6.19), we obtain

$$E = (2\pi A)^2 \gamma / d \qquad (6.46)$$

The effective mass per unit depth in the x-direction depends upon the effective depth in the z-direction. It seems reasonable to assume that the effective depth will be a function of the distance d and since the amplitude 2A of the ripple pattern is very shallow by comparison to d a reasonable expectation would be that the effective depth would be much less than d. Let us take the depth to be 'ad' where a< 1. The effective mass may then be written

$$M = a\rho d^2 \tag{6.47}$$

The amplitude of the displacement is taken to be

$$G = 2A \tag{6.48}$$

Thus, the natural frequency of oscillation is

$$f_r = [(2\pi A)^2 \gamma / a\rho d^3]^{1/2} / 4\pi A \tag{6.49}$$

In terms of the radian frequency, ω, and the spatial frequency, η, Eq. (6.49) becomes

$$\omega_r^2 = \gamma \eta^3 / 8\pi a\rho \tag{6.50}$$

The magnitude of 'a' can be determined by both experiment and by more rigorous analysis to be approximately $1/8\pi$ so that

$$\omega_r^2 = \gamma \eta^3 / \rho \tag{6.51}$$

6.4 EFFECTS PRODUCED BY PULSING THE SOUND WAVES

More precise analysis of the transient response of liquid surfaces have been carried out by Bander[2] and Pille[3]. In the latter analysis a solution to a boundary value problem defining the sound field in the imaging tank was obtained by numerical integration. It was demonstrated that an analytical expression could be formulated to fit the results obtained by numerical integration. Again substituting the symbol h for z in Eq. (6.15) we take note of the temporal characteristics by defining a function h (y,t) such that

$$h(y,t) = H(t)h(y)/2A \tag{6.52}$$

The analytical expression for $H(t)$ is characterized by two parameters ζ and ω_n. The damping ratio ζ measures how rapidly the oscillation decays and ω_n is the natural frequency of oscillation. For a radiation of duration Δt the response $H(t)$ can be written as

$$H(t)/2A = [1 - (1/\beta) \exp (-\zeta\omega_n t) \sin (\omega_d t + \theta)]U(t)$$
$$-[1-(1/\beta) \exp (-\zeta\omega_n(t - \Delta t)) \tag{6.53}$$
$$\sin (\omega_d(t - \Delta t) + \theta)]U(t - \Delta t)$$

where
$$U(t) = \begin{array}{l} 1 \text{ for } t > 0 \\ 0 \text{ otherwise} \end{array} \tag{6.54}$$

$$\beta = (1 - \zeta^2)^{1/2} \tag{6.55}$$

$$\omega_d = \beta \omega_n \tag{6.56}$$

and
$$\theta = \tan^{-1}(\beta/\zeta) \tag{6.57}$$

Agreement with numerical results is obtained if

$$\zeta = 2\eta\rho\nu \ [\eta/\rho(\rho g + \gamma\eta^2))]^{1/2} \tag{6.58}$$

and

$$\omega_n^2 = (\rho g + \gamma\eta^2)(\eta/\rho) \ \tanh \ (\eta D) \tag{6.59}$$

providing that D, the depth of the fluid in the imaging tank is greater than $\Lambda/4$. Furthermore, if $\rho g << \gamma\eta^2$ and $D>d/2$ as is normally the case

$$\omega_n^2 = \gamma\eta^3/\rho \tag{6.60}$$

which is identical to Eq. (6.51). In Eq. (6.58) ν is the kinematic viscosity.

REFERENCES

1.   H. P. Hildebrand and B. B. Brenden, An Introduction to Acoustical Holography, Plenum/Rosetta Edition, Plenum Press, New York (1972)

2.   Ref. 1 Sec. 6.5

3.   P. Pille and B. P. Hildebrand, Rigorous Analysis of the Liquid Surface Acoustical Holography System, in Acoustical Holography, Vol. 5, P. S. Green, Ed. Plenum Press, New York (1974) pp 335-372.

Chapter 7

THEORY ON IMAGING WITH ARRAYS

Albert Macovski

Stanford University

Stanford, California 94305

INTRODUCTION

A single ultrasonic transducer, of the type used in
commercial B-scan equipment, has a limited ability to
gather information. To accomplish many of the desirable
features, such as real-time operation, dynamic focusing
over a large depth range and electronic scanning, an array
must be used.

IMAGING ARRAYS

An array can be used as a near-field imaging structure
as shown in Fig. 1. In the far field of each transducer,

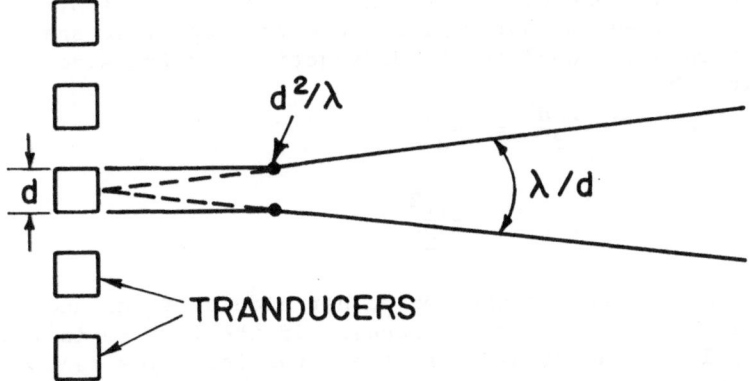

Fig. 1   Imaging Transducer Array Showing Diffraction
Spreading

111

at depths greater than d^2/λ, the angular pattern is
closely approximated by the spatial Fourier transform of
the transducer. This Fourier transform is a (sin x)/x
function for rectangular transducers and a $[J_1(x)]/x$

function for circular transducers. In each case the effective width of the angular pattern is approximately λ/d.
In the regions very close to the transducer the pattern
is simply that of the transducer output itself. This
pattern begins to distort at some distance before the plane
where the geometric projection intersects the far-field
pattern corresponding to d^2/λ. Thus, for high resolution
imaging, a near field structure of this type is adequate
only for relatively short distances. For example, using
3mm transducers and a wavelength of .5mm, corresponding
to 3.0 mHz, the near field distance is about 2 cm. This
distance is adequate for studies of the eye or of the
infant heart. For deeper body regions such as abdominal
or adult heart studies, the transducers must be made
significantly larger to avoid excessive divergence in the
far field regions.

One method of providing improved performance with
imaging arrays is the use of acoustic lenses. These have
the property of placing an image of the array at any
desired plane within the body. If the lens is placed in
the near field of the transducer beam, the size of the
pattern in the focal plane is given by $\lambda f/d$ where f
is the focal length. The depth of focus of the system
can be defined as that region where the geometric shadow
of the rays is equal to the diffraction limited size and
is given by

$$\frac{\triangle d}{f} = \frac{\lambda f}{d}$$

$$2 \triangle = \frac{2\lambda f^2}{d^2}$$

where $\triangle$ is the distance on either side of focus and thus
2 $\triangle$ is the total depth of focus. If the lens is in the
far field of the transducer, the diverging waves tend to
fill the lens aperture. The image of the transducer is
found in a plane determined by the lens law
$1/f = 1/d_i + 1/d_0$ where d_i and d_0 are the object and
image distances. The size of the image, for unity

magnification $(d_i = d_0)$, is given by the convolution of
the transducer pattern with the point response of the lens.
This point response is the Fourier transform of the lens
aperture and has an effective size of approximately $\lambda \, d_i / D$
where D is the diameter of the lens aperture.

   Imaging arrays can be operated in a number of modes.
It is convenient to use the same transducer array for both
transmitting the ultrasonic pulse and receiving the
resultant echoes. The round trip time of an ultrasonic
pulse using a 25 cm depth and a velocity of 1500 meters/sec
is about 333 μsec. Thus a real-time B-scan system can be
achieved using about 100 elements in sequence for a total
frame repetition time of 1/30 second. These 100 elements
can be arranged in a linear array, as shown in Fig. 1,
and provide a cross-sectional image of the reflections in
the plane of interest. If a lens system is used for
improved resolution, the round trip time will increase
because of the increased path length through an external
water bag containing the lens. Thus, for real time imaging,
the total number of elements must be reduced. One of the
difficulties in this configuration is the reflections from
the lens interfaces which are then re-reflected from the
transducer and appear as an artifact in the image. To
avoid this artifact the transducer array can be moved
about 25 cm from the lens. This approximately doubles
the round trip time and thus halves the number of transducers which can be used in a 1/30 second frame time. Anti-
reflection coatings can minimize the reflections at the lens
and the transducer and thus avoid this requirement.

   Imaging arrays can also be used in the C mode for
imaging a plane normal to the direction of propagation.
This has the desirable characteristic of having the same
prospective as that of conventional radiographs. However,
to accomplish C-scan imaging a two-dimensional array is
normally required. B-scan systems provide one dimension
by the propagation of the pulse so that only a line array
is required to image a cross-sectional plane. In C-scan,
both imaging dimensions are normal to the direction of
propagation.

   C-scan arrays can be used to study the reflection or
transmission properties of a planar region. For real-time
operation, in the reflection mode, the system is again

limited by the 333 μsec round trip time. If a 100 x 100 resolution image is desired, a sequential operation will take 3.3 seconds and thus negate real time operation. For real time each 100 element column or row can be excited simultaneously and used to simultaneously receive the echoes from a specific plane. A range gate defines the plane of interest and the resultant signals are stored and read out onto a line of the display while the next line is addressed. C-scans have the difficulty that an atten-uating structure in the beam path can shadow parts of the desired plane. This problem can be minimized by using a large aperture lens system where each point in the plane is addressed by a large-angle beam to minimize shadowing. A lens system also provides optimum resolution in the desired plane.

In a transmission imaging system an additional trans-ducer is used to insonify the plane from the opposite side of the region under study. The receiver array is focused onto the plane of interest using a lens. In this system range gating cannot be used to isolate a plane since the time of flight from the transmitter to the receiver trans-ducers is the same for all planes. Thus the lens imaging system of itself isolates the plane of interest. This is a problem because of the partial coherence of ultrasonic systems. Structures outside of the desired plane of focus cause undesired artifacts which interfere with the desired plane. These can be made blurred and less distracting by reducing the system coherence through broad bandwidth signals and spatially incoherent sources.

ARRAYS USING ELECTRONIC SCANNING AND FOCUSING

Imaging arrays have many limitations which do not fully exploit the full potential of ultrasonic imaging. Firstly, imaging systems do not provide the diffraction limited resolution capabilities over large depths. Thus the lateral resolution at any plane is limited by near field distance or depth of focus considerations. Secondly, imaging systems cannot conveniently generate a deflected beam which can generate a sector scan pattern. Controlled patterns allow a large area to be explored, in real time, using a relatively small array size. These methods also can avoid the various reflection and path-length problems of acoustic lenses.

To study the field patterns of arrays we make use of the point response of ultrasonic systems which relate the pressure at some point r to that in an x,y plane. Since the system is linear we can use the superposition integral as given by

$$p(r) = \frac{1}{j\lambda} \iint p(x,y) \, \frac{e^{jkr}}{r} \, ds$$

where $k = 2\pi/\lambda$, the wavenumber.

This formulation is identical to that used in Fourier optics to find the field at some point r due to a distribution on a surface. Using the same methods as that of Fourier optics we can derive the various Fresnel or near field and Fraunhofer or far field patterns. Assuming a source distribution in the $x_1, y_1, z = 0$ plane we can calculate the distribution at an observation plane x_0, y_0, z

where $r = \sqrt{z^2 + (x_0-x_1)^2 + (y_0-y_1)^2}$. For most systems, the paraxial approximation can be used for almost all regions except those very close to the x_1, y_1 plane. In that case, we use the first two terms of the binomial expansion as given by

$$r \cong z \left[1 + 1/2 \left(\frac{x_0-x_1}{z}\right)^2 + 1/2 \left(\frac{y_0-y_1}{z}\right)^2 \right]$$

The r in the denominator is replaced by z with good accuracy because of the relative insensitivity of this term. The r in the exponent is replaced by the approximate binomial expansion of r giving the pressure in the Fresnel region as

$$p(x_0,y_0) = \frac{e^{jkz}}{j\lambda z} \iint p(x_1,y_1) \exp \left\{ j \frac{k}{2z} \left[(x_0-x_1)^2 + (y_0-y_1)^2 \right] \right\}$$

$$\cdot \, dx_1 dy_1$$

Unfortunately this integration is often quite involved because of the quadratic phase factors.

For further distances from the source, in the region where

$$z \gg \frac{kr_{1max}^2}{2}$$

the Fraunhofer approximation can be used where the terms involving x_1^2 and y_1^2 can be neglected. Thus in the far-field or Fraunhofer region

$$p(x_0,y_0) = \frac{e^{jkz}}{j\lambda} e^{j\frac{k}{2z} r_0^2} \iint p(x_1,y_1)$$

$$\cdot \exp\left[-j\frac{2\pi}{\lambda z}(x_0 x_1 + y_0 y_1)\right] dx_1 dy_1$$

Thus the pressure distribution in the Fraunhoffer region, except for a quadratic phase factor, is simply the Fourier transform of the pressure distribution at the x_0, y_0 plane where the spatial frequency variables are $x_0/\lambda z$ and $y_0/\lambda z$. The $e^{jkz}/j\lambda$ is simply a constant phase shift weighting factor.

Using these basic relationships we can calculate the pattern of arrays. Consider the transducer array shown in Fig. 2. The array shown has a pressure pattern given by

$$p(x_1,y_1) = rect\frac{x_1}{X}\left[comb\frac{x_1}{S} * rect\frac{x_1}{W}\right] rect\frac{y_1}{Y}$$

where $comb\ x = \sum_{n=-\infty}^{\infty} \delta(x-n)$. This is simply a representation of the array of rectangular transducers of dimensions $W \times Y$ separated by S. The far field pressure pattern, as before is a quadratic phase factor multiplying the Fourier transform of the source pattern as given by

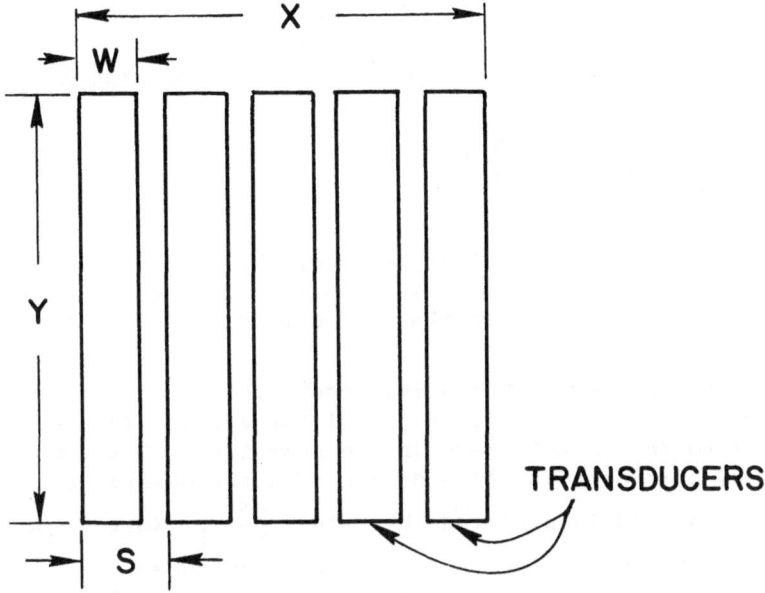

Fig. 2   Rectangular Transducer Array

$$p(x_0, y_0) = \frac{e^{jkz} e^{j\frac{k}{2z} r_0^2}}{j\lambda z} \left\{ X \operatorname{sinc}\left(\frac{x_0 x}{\lambda z}\right) \right.$$

$$\left. * \left[S \operatorname{comb}\left(\frac{S x_0}{\lambda z}\right) W \operatorname{sinc}\left(\frac{x_0 W}{\lambda z}\right) \right] \right\} Y \operatorname{sinc}\left(\frac{y_0 Y}{\lambda z}\right)$$

Using the relationship

$$S \operatorname{comb}\frac{S x_0}{\lambda z} = \sum_{n=-\infty}^{\infty} \delta\left(\frac{x_0}{\lambda z} - \frac{n}{S}\right)$$

we obtain

$$p(x_0, y_0) = \frac{e^{jkz} e^{j\frac{k}{2z} r_0^2}}{j\lambda z} \, WXY \left\{ \sum_{n=-\infty}^{\infty} \mathrm{sinc}\left(\frac{nW}{S}\right) \, \mathrm{sinc} \, X\left(\frac{x_0}{\lambda z} - \frac{n}{S}\right) \right\}$$

$$\cdot \, \mathrm{sinc}\left(\frac{y_0 Y}{\lambda z}\right)$$

In the y_0 dimension this is identical to the far field pattern of the rectangular aperture and has an angle of approximately λ/Y. In the x_0 direction the pattern is given in Fig. 3.

This pattern shows a main central lobe whose width is determined by the width X of the array. This corresponds to the $n = 0$ term in the summation. The remaining terms, corresponding to side lobes of the desired pattern, are modulated in amplitude by $\mathrm{sinc} \, \dfrac{nW}{S}$. Notice that if $W = S$, corresponding to contiguous transducer elements in the array, the side lobes disappear since $\mathrm{sinc} \, n = 0$ for $n \neq 0$. If W is an appreciable percentage of S, as in the case illustrated in Fig. 3, the side lobes become significantly attenuated.

The sole purpose of this array is to provide a mechanism

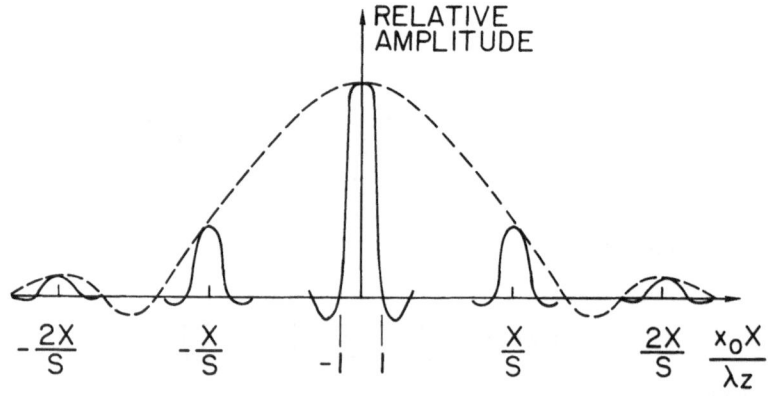

Fig. 3   Far Field Pattern of a Rectangular Array

for controlling the array pattern. One important control mechanism is deflection. The beam can be steered in a sector scan pattern by providing delays or phase shifts to each element in the array proportional to x_1. The source pattern then becomes

$$p(x_1,y_1) \;=\; \text{rect } \frac{x_1}{X} \left[\left(\text{comb } \frac{x_1}{S} \right) e^{j\beta x_1} * \text{rect } \frac{x_1}{W} \right] \text{rect } \frac{y_1}{Y}$$

where each element is given a phase shift of β n S. The multiplication of comb $\dfrac{x_1}{S}$ by $e^{j\beta x_1}$ in the far field or Fourier transform region corresponds to the convolution of

S comb $\left(\dfrac{Sx_0}{\lambda z} \right)$ with $\delta\left(\dfrac{x_0}{\lambda z} - \dfrac{\beta}{2\pi} \right)$ which provides

$$S \text{ comb } \left(\frac{Sx_0}{\lambda z} \right) * \delta\left(\frac{x_0}{\lambda z} - \frac{\beta}{2\pi} \right) \;=\; \sum_{n=-\infty}^{\infty} \delta\left(\frac{x_0}{\lambda z} - \frac{n}{S} - \frac{\beta}{2\pi} \right)$$

Using this operation, the resultant far field pattern is given by

$$p(x_0,y_0) \;=\; \frac{e^{jkz} e^{j\frac{k}{2z} r_0^2}}{j\lambda z} \; WXY \left\{ \sum_{n=-\infty}^{\infty} \text{sinc } W \left(\frac{n}{S} + \frac{\beta}{2\pi} \right) \right.$$

$$\left. \text{sinc}\left[X\left(\frac{x_0}{\lambda z} - \frac{n}{S} - \frac{\beta}{2\pi} \right) \right] \right\} \; \text{sinc } \left(\frac{y_0 Y}{\lambda z} \right)$$

This has the same pattern in the y_0 direction and a pattern in the x_0 direction shown in Fig. 4. Note that the main lobe has been deflected in the x_0 plane. Using x_0/z as the approximate angle, the beam has been deflected an amount $\dfrac{\lambda\beta}{2\pi}$. The important undesired effect is the increased relative amplitude of one of the sidelobes. The sum of sinc patterns is modulated by the weighting factor sinc $W\left(\dfrac{n}{S} + \dfrac{\beta}{2\pi} \right)$. With deflection, the desired $n = 0$ term is reduced from unity to sinc $W \left(\dfrac{\beta}{2\pi} \right)$ while the undesired

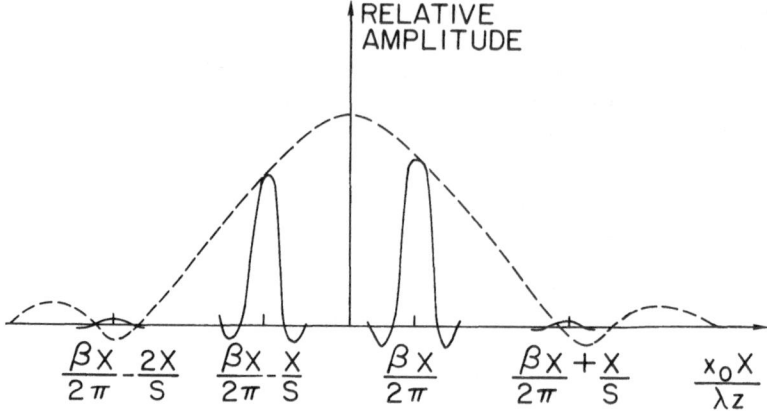

Fig. 4   Far Field Pattern of a Rectangular Array with
         Deflection

$n = -1$ term is increased to sinc $W\left(\dfrac{\beta}{2\pi} - \dfrac{1}{S}\right)$. Thus, even
using contiguous transducers, where $W = S$, the side lobe
remains a problem. While receiving energy from the angle
$\dfrac{\lambda\beta}{2\pi}$, the array will also be receiving energy from an angle
of $\dfrac{\beta\lambda}{2\pi} - \dfrac{\lambda}{S}$ which can reduce the directivity of the system
and provide false responses. One solution to this problem
is the use of limited deflection angles. This is often
undesirable for clinical applications. For example in
cardiac studies a deflection angle of approximately 90°
is often desired to visualize most of the heart. Another
solution is the use of an insonification array system which
doesn't illuminate the side lobe pattern of the receiver
array system. Thus if the transmitter array and receiver
array have identical main lobes but dissimilar positions
of their side lobes, the problem can be minimized. The use
of large numbers of elements, where x/S is a relatively
large number, will reduce the side lobes at the price of
increased complexity.

The previous results relate exclusively to the far
field. In the near field the patterns represent the
Fourier transform of the source pattern multiplied by the
quadratic phase factor $\exp j(k+2z)[x_1^2 + y_1^2]$. This is

equivalent to convolution of the Fourier transformed far field pattern by $j\lambda z \exp-(j \pi/\lambda z)[x_0^2 + y_0^2]$. As before, this has the effect of creating a pattern which starts out looking like the source at low values of z and then becomes oscillatory until it evolves into the Fourier transform patterns where the quadratic term becomes negligible. In order to achieve diffraction limited resolution in the near field region the array pattern can be focused with a lens in the x_1,y_1 plane. The lens has a transmission given by $\exp-j(k/2f)[x_1^2 + y_1^2]$. The focal length is given by

$$f = \frac{R}{n_1 - n_2}$$

where R is the radius of curvature of the lens surface and $n_1 - n_2$ is the difference in refractive index between the lens material and its surround. At the depth plane where $f = z$ the quadratic phase factor will be cancelled and the system will achieve the desired diffraction limited resolution.

As with deflection, it is desired to also electrically control the distance to the focal plane. Rather than using phase shift proportional to distance x_1, as was done for deflection, a phase shift is added at each element proportional to x_1^2. The resultant source pattern becomes

$$p(x_1,y_1) = \text{rect } \frac{x_1}{X} \left[\left(\text{comb } \frac{x_1}{S}\right) e^{-j\gamma x_1^2} * \text{rect } \frac{x_1}{W} \right] \text{rect } \frac{y_1}{Y}$$

In the near field this pattern becomes

$$p(x_0,y_0) = \frac{e^{jkz} e^{j\frac{k}{2z} r_0^2}}{j\lambda z} \; \mathcal{F}[\exp j \left(k/2z\right)(x_1^2 + y_1^2)p(x_1,y_1)]$$

where $\mathcal{F}$ is the Fourier transform operator. The focusing effect can be seen from the following relationship

$$e^{j\left(k/2z\right)x_1^2}\left[\left(\text{comb}\ \frac{x_1}{S}\right)c^{-j\gamma x_1^2}*\text{rect}\ \frac{x_1}{W}\right]$$

$$= S\sum_{n=-\infty}^{\infty}\frac{\text{rect}\left(x_1 - nS\right)}{W}e^{j\left[\frac{k}{2z}x_1^2 - \gamma(nS)^2\right]}$$

This relationship indicates the extent of the quadratic phase shift at each element n due to diffraction and the phase shifter γx_1^2 added to each element. At $\gamma = \frac{k}{2z}$, the phase shift in the center of each element is zero. Thus the array itself no longer has an overall quadratic phase factor relative to depth plane z. Each individual element continues to have a small quadratic phase factor over the distance W. However, this can be neglected in the far field region of W. For relatively thin elements in the array, the far field of each element is essentially all depths of interest. Thus, when using quadratic phase shifts at each element the pattern at distance z, where $\gamma = k/2z$, is essentially identical to the far field pattern previously derived using the Fraunhofer approximation. The value of γ can be adjusted for any desired depth plane. In pulsed systems, γ can be time varying and thus follow a propagating pulse. To provide deflection plus focusing linear phase shifts are superimposed on the quadratic phase shifts. The system shown provides controlled deflection and focusing in the x dimension only. A two-dimensional array of transducers, with the appropriate linear and quadratic phase shifts applied to the elements can provide two-dimensional focusing and deflection to any point in space.

SYSTEMS PROVIDING TWO-DIMENSIONAL FOCUSING

The use of arrays to accomplish deflection and real-time focusing in both dimensions requires significant complexity. In general, a relatively broad illumination pattern is used in the transmit mode since dynamic focus can only be accomplished in the receive mode. This transmitter is steered using linear phase shifters. On receive, each of the n x n transducer elements in a two-dimensional array requires an independently controlled time-varying delay line to provide dynamic focus in both dimensions. Thus to achieve a B-scan of n lines, approximately n^2 transducers and time-varying delay lines are required.

One method of simplification is to have the desired field pattern be the product of individual transmitter and receiver array patterns so that each can provide focus in one dimension. However, since dynamic focus can only be accomplished on a receiver array, a transmitter array pattern must be chosen which provides the desired performance at all depths of interest. This is quite straightforward in the C-scan mode. The transmitter array pattern has the proper quadratic phase shifts to focus at the plane of interest. One simple example of this is a cross configuration where one arm is used to transmit and the other to receive. Using the appropriate delays, a line in the plane is illuminated with the transmit arm. The receiver arm is also focused on the same plane with various linear delay networks used to simultaneously view each element in the line. The transmitter beam is successively deflected through each line in the plane so that an $n \times n$ image is formed in n round-trip intervals. This system provides a C-scan in real time and requires delay elements which are switched following each propagation interval. No delay elements are varied during the propagation interval since C-scan systems do not require dynamic focus.

The method described cannot be used for B-scan systems since the transmitted beam can only be focused in a single plane. Rather than use the complex $n \times n$ array processing, we can make use of the interesting properties of circular arrays. These have the same relative field patterns at all depths since the Fraunhoffer and Fresnel regions have the same amplitude characteristic. In general an annular ring having any type of angular functions around the ring can be expressed as summations of the form

$U_n(r, \theta) = \delta(r_1 - R)e^{jn\theta}$ where r_1 is the radial variable

at the array and R is the radius of the annulus. The far field or Fraunhoffer pattern is given by

$$U_n(r, \phi) = C \exp\left(j \frac{\pi}{\lambda z} r^2\right) \mathfrak{F}[U(r_1, \theta)]$$

$$= 2\pi R C (-j)^n e^{jn\phi} \exp\left(j\frac{\pi}{\lambda z} r^2\right) J_n\left(\frac{kRr}{z}\right)$$

where $\mathfrak{F}$ is the Fourier transform operator. The near field or Fresnel pattern is given by

$$U_n(r,\phi) = 2\pi RC(-j)^n e^{jn\phi} \exp\left(j\frac{\pi}{\lambda z}r^2\right) \mathfrak{F}[U(r_1,\theta)\exp(j\frac{k}{2z}r_1^2)]$$

$$= 2\pi RC(-j)^n e^{jn\phi} \exp[j\frac{\pi}{\lambda z}(r^2 + R^2)]J_n\left(\frac{kRr}{z}\right)$$

Thus, except for a phase factor, the amplitude patterns are the same and are both characterized by the Bessel function $J_n\left(\frac{kRr}{z}\right)$. A uniformly weighted annulus, with $n = 0$, will provide a $J_0\left(\frac{kRr}{z}\right)$ pattern at all depths z. Although this pattern has a narrow central lobe, it has excessive side lobes and would thus be unsuitable.

The desired pattern can be realized by weighting the annulus with $\cos^2\theta$ which provides $e^{jn\theta}$ components at $n = 0$, -2, and $+2$. Ignoring constants and phase factors, this weighting provides a field pattern at all depths given by

$$U(r,\phi,z) = \frac{1}{2}\left[J_0\left(\frac{kRr}{z}\right) - J_2\left(\frac{kRr}{z}\right)\cos 2\phi\right]$$

Using the Bessel function recursion identity

$$J_{n-1}(x) + J_{n+1}(x) = \frac{2nJ_n(x)}{x}$$

the field can be expressed as

$$U(r,\phi,z) = \frac{J_1\left(\frac{kRr}{z}\right)}{\left(\frac{kRr}{z}\right)} - J_2\left(\frac{kRr}{z}\right)\cos^2\phi$$

In this form we see that the pattern in the y direction, where $\cos\phi = 0$, is $J_1(\cdot)/(\cdot)$. This is the classical diffraction limited resolution of a full circular aperture. Thus, in the y direction, the field pattern is identical to that of a lens which is focused at every depth plane z. This configuration thus achieves the desired transmitter pattern at all depths without dynamic focus.

This pattern has excessive side lobes in the x
direction, but this is taken into account by the receiver
array as shown in Fig. 5. The receiver array within
the annular transmitter array is the linear array previously
analyzed. Through dynamic focusing, using a time delay
proportional to x^2 , the array is focused at each depth

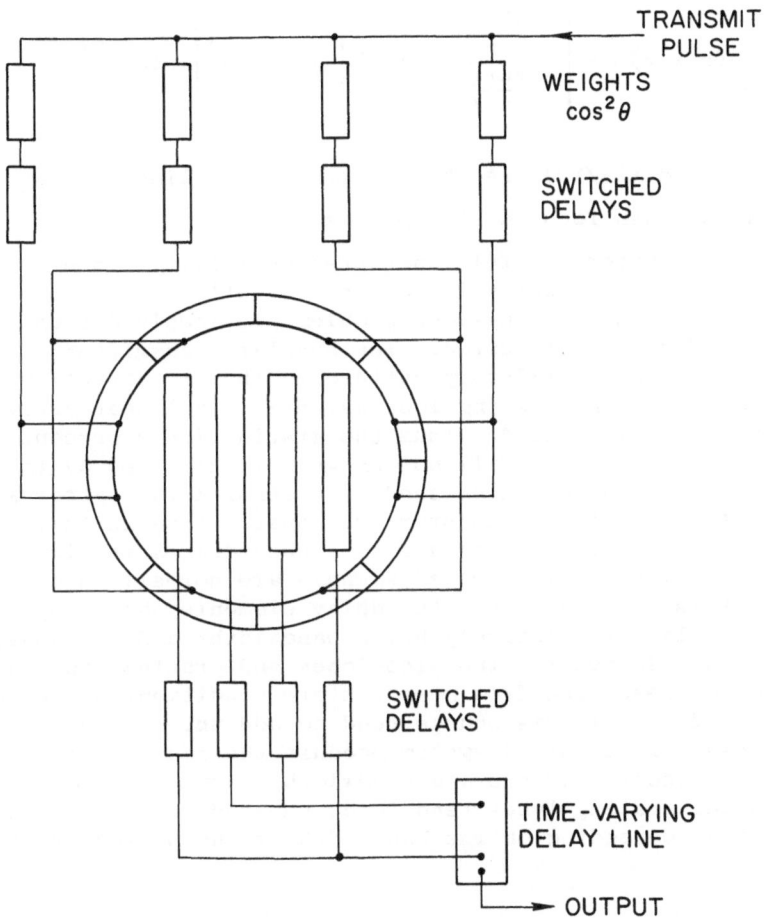

Fig. 5   Sector Scan System Using a Weighted Ring Array
          Transmitter and a Linear Array of Line Segments
          for the Receiver.

plane in sequence. For deflection, both the transmitter
and receiver arrays are deflected in the x direction
using a set of switched delays proportional to x which
are changed following each scan line.

The overall system of Fig. 5 forms a high performance
B-scan structure generating a real-time sector scan.
Through the properties of the weighted angular ring and the
dynamically focused linear array, diffraction limited
performance is achieved at all depths with a pattern given
by

$$U(r,x,z) = \left[\frac{J_1\left(\frac{kRr}{z}\right)}{\left(\frac{kRr}{z}\right)} - J_2\left(\frac{kRr}{z}\right) (x/r)^2 \right] \text{sinc} \frac{Xx}{\lambda z}$$

where   sinc (Xx/λz)   is the pattern of the linear array with
width   X   when focused at depth   z.

This system has full aperture sensitivity since the
receiver array essentially covers the entire aperture.
In addition the grating-lobe problem previously discussed
is considerably attenuated. The annular transmitter
pattern has an oscillatory pattern in the x direction but
does not have the grating lobe pattern of a linear array as
shown in Figs. 3 and 4. Thus the grating lobes present in
the receiver pattern will not be well-illuminated by the
annulus and thus be suppressed. In general the patterns
shown throughout this chapter are conservative in that they
are the steady state response to sinusoidal excitation.
In reflection imaging pulsed signals are normally used.
Thus it is important that the delay elements shown represent
actual delay at relatively broad bandwidths and not phase
shifters. In general the side-lobes and grating lobes are
phenomena resulting from steady state conditions where the
phases of an off-axis source tend to add up. These are
suppressed in a pulsed system because the additive nature
does not occur over a broad bandwidth. Thus the side-
lobes and grating lobes tend to be smeared out and of reduced
amplitude depending on the bandwidth of the pulsed excitation.

Chapter 8

INTEGRATED ELECTRONICS FOR ACOUSTIC IMAGING ARRAYS

James D. Meindl

Department of Electrical Engineering
Stanford University
Stanford, CA 94305

ABSTRACT

Within the past five years acoustic imaging systems based on the use of arrays of transducer elements rather than a single element have undergone a rapid stage of development. Primary performance objectives of these systems in medical applications include high resolution, large field of view, long range and rapid frame rate. Various approaches to array systems can be classified according to transducer geometry, scanning technique and method of focus. Analysis of the salient electronic device and circuit requirements of these systems indicates well matched circuit arrays of 1) high voltage transmitters, 2) low noise, wide dynamic range preamplifiers and logarithmic compressors, 3) high voltage multiplex switches and 4) electronically variable analog delay lines are needed, with variable delay lines representing perhaps the most acute requirement. Custom silicon monolithic integrated circuits offer a promising means for fulfilling these needs, particularly in the case of charge coupled device (CCD) delay lines. Array imaging systems offer promise of extraordinary future progress.

I. INTRODUCTION

Within the past five years acoustic imaging systems based on the use of arrays of transducer elements [1,2] rather

than a single element [3] have undergone a rapid stage of
development which may well persist in the foreseeable future.
A generalized block diagram of such a system is illustrated in
Fig. I-1. Here the "object" represents the structure to be
imaged. The "transducer" consists of an array of one or more
acousto-electric elements. The "preprocessor" typically
performs on-line real-time analog signal processing. The
"processor" usually provides programmable off-line digital
signal processing and computation. And, the "display"
performs the final electro-optical image conversion.

The principal objective of this discussion is a review
of the salient electronic device and circuit advances which
have served to implement the recent development of acoustic
imaging arrays. From this review, which naturally focuses
on preprocessor (or array) electronics, one seeks an appreciation of the more promising opportunities for future progress
in acoustic imaging systems.

Since the performance objectives for array electronics
invariably represent a compromise between system requirements on one side and technological capabilities and economic
resources on the other, Section II of this paper reviews the
primary requirements--resolution, field of view, range and

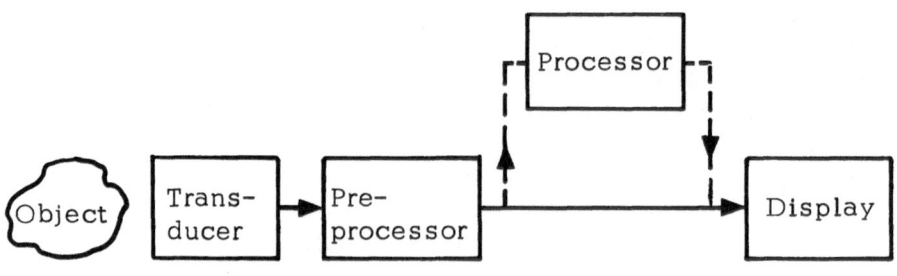

Fig. I-1. Generalized block diagram of acoustic imaging system

frame rate--of acoustic imaging systems for medical applications. Although somewhat restrictive, narrowing the system requirements to fit this particular class of applications offers the great benefit of more concrete illustration of key system design trade-offs. In addition, most imaging array effort which has been reported to date has been concerned with medical applications.

In order to facilitate the organization of following sections of the discussion, Section II also presents a simple classification of array imaging systems based upon transducer geometry, scanning technique and method of focus.

Because of their central importance in array electronics, Section III emphasizes the primary features of receivers and transmitters for single transducer imaging systems. Section IV reviews various types of linear array systems which have been reported and underscores the importance of electronic delay lines and multiplex switches in these systems. Area array systems are considered in Section V. Here, a substantial need for new transducer interconnection techniques is added to array electronics requirements.

Throughout this discussion a special effort is made to point out the advantages which may accrue to an acoustic imaging system as a result of the utilization of custom integrated electronics. It has been demonstrated repeatedly in the largest computer and communication systems as well as in the smallest electronic wrist watches and pocket calculators that substantial benefits can be derived from customizing the design of silicon monolithic integrated circuits to fulfill the peculiar needs of a specific system or class of systems. The imposing electronics requirements of acoustic imaging arrays offer a multitude of extremely promising opportunities-- for example, in receivers, transmitters, delay lines, multiplex switches and interconnections--to exploit the unique potential of custom integrated electronics [4].

Although this discussion is limited to acoustic imaging systems which incorporate arrays of electronic transducers for the acoustic-to-electrical conversion process, one should not infer therefrom that other promising approaches do not exist. In fact, the variety of approaches to acoustic imaging now under investigation is exceedingly remarkable [1,2]. A principal reason for restricting this discussion to electronic transducers is that their sensitivity is generally about 100 times better than that of alternative approaches such as the "static and dynamic ripple" techniques which employ optical detection of surface deformations [5]. This sensitivity advantage is of primary importance in medical applications.

II. SYSTEM REQUIREMENTS

The principal performance requirements of an acoustic imaging system intended for medical application are listed in Fig. II-1. A brief review of these requirements is presented in this section.

SYSTEM REQUIREMENTS

A. RESOLUTION
   1. HIGH
   2. AXIAL, LATERAL
   3. 1D, 2D, 3D (Volumetric)

B. FIELD OF VIEW
   1. LARGE
   2. 1D, 2D, 3D (Volumetric)
   3. REFLECTION, TRANSMISSION

C. RANGE
   1. LONG
   2. RECEIVER SENSITIVITY, DYNAMIC RANGE
   3. TRANSMITTER POWER, FREQUENCY

D. FRAME RATE
   1. RAPID
   2. REAL TIME

Fig. II-1. System requirements of an acoustic imaging system intended for medical application

A. Resolution

Perhaps the most basic function of an imaging system per se is the formation of an image with sufficient fidelity to convey useful information regarding the spatial characteristics of a corresponding object. To this end "high" resolution or the ability to resolve fine spatial dimensions is a key performance requirement of medical ultrasonic imaging systems. In practice, it is of particular importance in such systems to distinguish between axial resolution and lateral resolution. For example, in a collimated single transducer pulse-echo or reflection system, axial resolution refers to resolving power along the linear path traveled by a transmit pulse. It may be as little as 1.0 mm. However, the corresponding lateral resolution, i.e. resolving power orthogonal to the transmit pulse axis, is typically more than 1.0 cm [1]. Clearly, one must beware of this order of magnitude difference in system "resolution." In addition, substantial variations in axial resolution as a function of transmit power and in lateral resolution as a function of range are not uncommon in both single transducer and linear array systems [1,2]. Moreover, in linear array systems which utilize electronic sector scanning, lateral resolution in the plane of the scan may be several times better than lateral resolution orthogonal to the scan plane (e.g., 3-4 mm vs 9-12 mm) [3]. Finally, it is important to recognize that anatomic organs are three-dimensional objects of irregular shape and arbitrary orientation with respect to an acoustic transducer. Consequently, unambiguous measurements of their dimensions can be assumed only to the degree of the three-dimensional or "volumetric" resolution of the imaging system. That is, in general, high resolution images of an organ can be achieved only if the volumetric resolving power of the imaging system is high in all three spatial dimensions.

B. Field of View

The field of view of medical imaging systems should be large in order to permit a single display of the entire region subject to examination and to provide a useful perspective

view of an organ of particular interest (e.g., an entire
heart contained within a 10 x 10 x 10 cm volume). Basically, a single transducer pulse-echo imaging system is
restricted to a one-dimensional (1D) field of view which
consists of the line of travel of a transmit pulse. Through
manual or mechanical scanning, this 1D field is often
extended to a plane [1]. A two-dimensional (2D) field of
view is characteristic of linear array systems [2,3]. Area
arrays may offer the most promising opportunities for
achieving the full three-dimensional (3D) or volumetric
fields of view (with high resolution, real-time imaging
throughout the volumetric fields) which are necessary for
maximum effectiveness in medical imaging [4]. An additional factor which can markedly influence the field of
view of a medical imaging system is its basic mode of
operation. A reflection system with a transmit/receive
transducer array inherently makes optimum use of the
limited set of apertures the body provides, thus enhancing
both resolution and field of view. In contrast, transmission
imaging of the heart, for example, is rather difficult.

C. Range

    The range of a medical imaging system must be
sufficient to permit examination of relatively deep-lying
organs as well as more superficial structures. For example,
a range of 25 cm or more is desirable for abdominal imaging. Tissue absorption, specular reflection and small
changes in tissue acoustic impedances are primary causes
of signal attenuation in pulse-echo systems. High sensitivity receivers (with wide dynamic range), high transmit
powers and low excitation frequencies are useful means of
extending range. However, signal sensitivity is limited
by transducer noise, transmit power by patient safety considerations, and excitation frequency (or wavelength) by
the required resolution. Consequently, useful operating
range remains an important parameter in the design trade-
offs of medical imaging systems.

D. Frame Rate

The frame rate of an imaging system should be rapid enough to resolve object motion. This permits movement of the imaging transducer array with respect to an object in order to resolve ambiguities related to the three-dimensional nature of the anatomy. In addition, it helps an observer to disregard many system artifacts. And, most importantly, it permits accurate imaging of a moving target such as a fetus and precise studies of target dynamics. For example, studies of cardiac motion are possible with frame rates of 15-30 frames per second, and accurate dynamic measurements of heart chamber volumes are quite feasible with the volumetric frame provided by an area array system [4].

E. Additional Factors

The system requirements summarized in Fig. II-1 reflect the demands which might be imposed on a system intended for real-time morphological imaging. It is well to recognize that additional quantitative measurements of such tissue parameters as attenuation, acoustic velocity, acoustic impedance and echo-scattering cross section may be of great significance [5]. Given an imaging system with high volumetric resolution, a large volumetric field of view, long range and rapid frame rate, one may well possess an excellent tool for investigation of these additional factors.

F. Classification Scheme

Numerous trade-offs between resolution, field of view, range and frame rate are present in the design of an ultrasonic imaging system. The specific nature of these trade-offs varies markedly with the overall design approach which is assumed. Fig. II-2 presents a classification scheme which encompasses most of the approaches to array systems which have been reported to date. It is apparent that this scheme is based upon transducer geometry, scanning technique and method of focus. It serves as the basis for the organization of the following sections of this discussion.

CLASSIFICATION OF ARRAY SYSTEMS

I.   Single Transducer Systems
      A. Manual Scan
           1) Unfocused
           2) Acoustic Focus
      B. Mechanical Scan

II.   Linear Array Systems
      A. Manual Scan
      B. Mechanical Scan
           1) Acoustic Focus
           2) Electronic Focus
           3) Holographic
      C. Electronic Scan
           1) Rectilinear Scan
           2) Sector Scan
                a) Unfocused
                b) Cylindrical Focus
           3) Orthogonal Scan
           4) Holographic

III. Area Array Systems
      A. Acoustic Focus
      B. Electronic Focus
      C. Holographic

Fig. II-2.   Classification of array systems

III.   SINGLE TRANSDUCER SYSTEMS

     Chiefly because of its widely demonstrated utility, the simple single transducer pulse-echo imaging system offers an appropriate point of departure for reviewing the properties of acoustic imaging arrays for medical diagnosis. In this section strong emphasis is placed on receiver and transmitter properties because of their central importance in array systems.

A. Manual Scan

1) Unfocused

A simplified block diagram of a single transducer system is illustrated in Fig. III-1. The central timing function or clocking of the system is performed by the rate generator which is, in essence, an astable multivibrator oscillating at a fundamental frequency of several kilohertz or less. Each output pulse from the rate generator causes the transmitter to excite the transducer with a short ($\sim 1\mu$sec) burst of high voltage sinusoidal excitation typically with a frequency in the 1-5 MHz range. Simultaneously the rate generator triggers a) the sweep generator to initiate the horizontal sweep of the cathode ray tube (CRT) display and b) the swept gain generator to begin to increase the RF amplifier gain as a function of the distance of the transmit pulse from the transducer in order to compensate for tissue absorption. Ultrasonic echoes returning from the target are received by the transducer and subjected to further processing in the RF amplifier detector and suppressor prior to their presentation in an amplitude versus range (A-scan) or intensity versus position (B-scan) format on the CRT display.

The basic principle of the A-scan (or A-scope) as described by Wells [1] is illustrated in Fig. III-2. A piezoelectric transducer (probe) in direct contact with the skin emits a

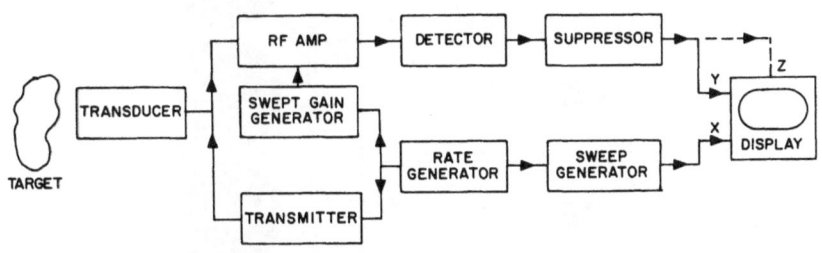

Fig. III-1.   Single transducer pulse-echo ultrasonic imaging system

pulse of ultrasonic energy into the body (medium i) and simultaneously triggers the horizontal (x-axis) sweep of the CRT
display. In essence, the signal appearing across the transducer electrodes is applied to the vertical (y-axis) deflection
plates of the CRT so that the pulse tracing first appearing in
Fig. III-2b & c represents the transmit signal, while the added
pulse appearing in Fig. III-2f represents an ultrasonic echo
returning from the interface between media i and ii, which
differ in their acoustic properties. Assuming, for example,
that the acoustic velocity in medium i is known, the spacing
of the two pulses in Fig. III-2f can be used to define the thickness of medium i along the line of travel of the ultrasonic pulse.

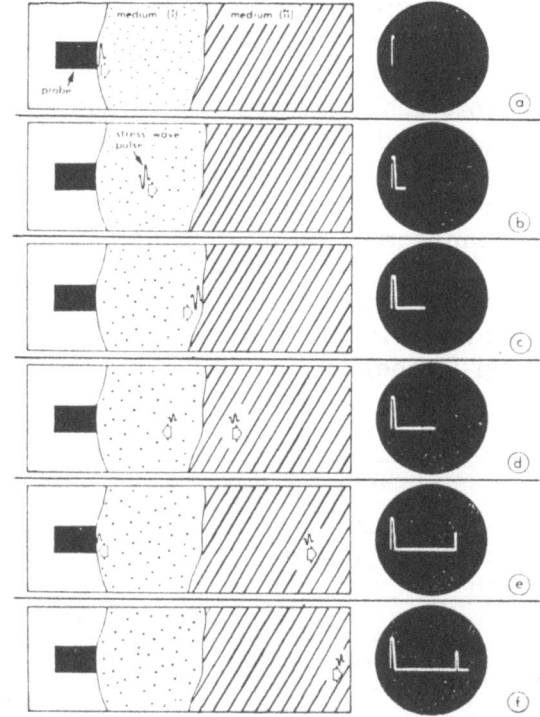

Fig. III-2.   Basic principles of the pulse-
echo system (after Wells)

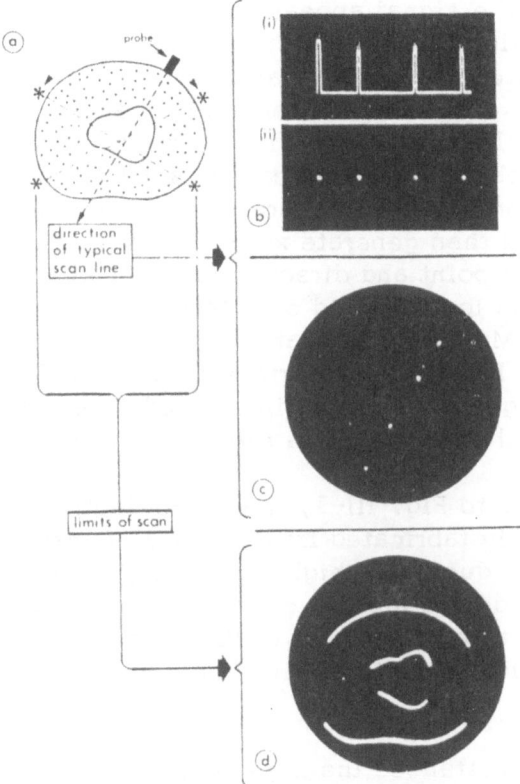

Fig. III-3. A-scope and B-scope methods for displaying
            ultrasonic pulse-echo information. (a) Schematic
            representation of a section through a patient;
            (b) (i) A-scope presentation of a typical scan line;
            (ii) B-scope presentation of the same scan line;
            (c) B-scan as in (b)(ii), but with the direction of
            the timebase linked to the direction of the ultra-
            sonic beam; (d) compound B-scan, integrated
            from many individual scans, each one similar to
            (c). (after Wells)

The A-scan mode of operation illustrated in Figs. III-2
and III-3b(i) provides an amplitude versus range display.
Alternatively the signal appearing across the transducer electrodes is applied to the electron source (z-axis) of the CRT,
and the amplitude of this signal thus modulates the brightness
of the display as illustrated for the B-scan (or B-scope) display of Fig. III-3b(ii). In addition, the location and orientation of the probe on the body can be used to control potentiometers incorporated in a mechanical scanning arm. The
potentiometers then generate x- and y-axis input signals such
that the initial point and direction of the CRT beam sweep are
linked to probe location and orientation as illustrated in
Fig. III-3c. Manually, repeating this type of B-scan for
various probe locations and orientations provides a cross-
sectional image of the object integrated by a storage display
from individual scans as illustrated in Fig. III-3d.

Referring to Fig. III-1, the piezoelectric transducer
itself is usually fabricated from a wafer of lead zirconate
titanate (PZT), due to its high sensitivity. Thin metal electrodes are bonded to the flat surfaces of the wafer, which is
provided with a backing layer or block to dampen vibrations
and thus provide short transmit pulses and high axial resolution [1].

In most instances the high voltage transmitter circuit
provides shock or impulse excitation to the transducer [1],
as illustrated in Fig. III-4a [2]. In this circuit the control
transistor Q is normally saturated. When it is cut off by a
pulse from the rate generator, the trigger capacitor C_T is
charged by the low voltage (-10V) supply through the control
gate electrode of the silicon controlled rectifier (SCR). This
causes the SCR to turn on, thereby placing the shock excitation capacitor C_S which is charged by the high voltage supply
to 400V directly in parallel with the transducer, causing it to
emit a transmit pulse at its natural frequency. The damping
resistor R_D serves to control the level of the transmit pulse.
Although it permits relatively short excitation pulses, the
circuit of Fig. III-4a is rather inefficient. As illustrated in
Fig. III-4b, excitation of the transducer from a gated AC source

operating at the ultrasonic frequency provides improved efficiency and permits the use of lower voltage devices in the transmit circuitry. Although impulse excitation is generally preferable for single transducer systems, this may or may not be the case for array systems, depending on specific features of a design.

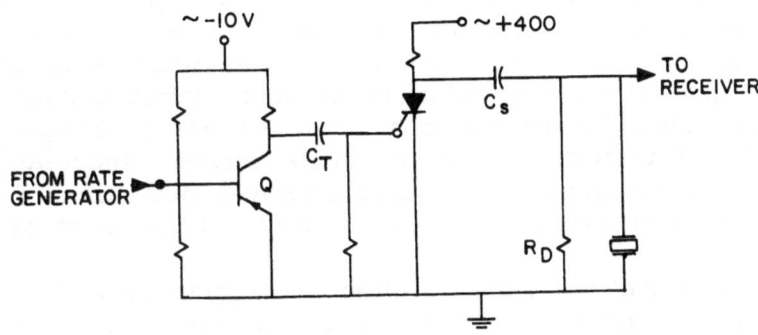

Fig. III-4a. Shock transmitter circuit

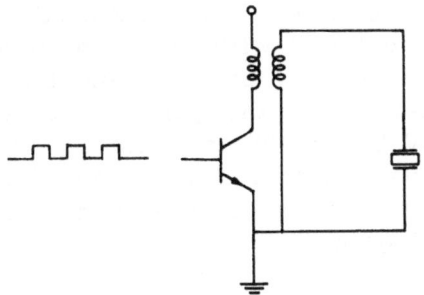

Fig. III-4b. Gated oscillator transmitter circuit

The receiver channel consisting of the RF amplifier, detector, suppressor and associated circuits is the key section of a single transducer pulse-echo ultrasonic imaging system. A graphic summary prepared by Wells [3] of the signal processing which may occur in the receiver is illustrated in Fig. III-5. Note that the swept gain amplification "brings up" echoes from more distant targets and thus tends to compensate for tissue absorption. Logarithmic transformation or compression is perhaps the most important post-detection process (and may also be performed as a pre-detection process) because it permits simultaneous display of signals of very wide dynamic range on a gray scale display of much more limited dynamic range. Thus logarithmic compression tends to compensate for the differences in signal levels between specular and diffuse reflections. It permits improved definition of convoluted boundaries along with internal structure of organs.

A somewhat more detailed block diagram of the RF amplifier portion of a pulse-echo receiver is illustrated in Fig. III-6. The principal performance requirements for the RF amplifier are high sensitivity (or low noise figure) and large dynamic range (via swept gain and gain compression). A block diagram of an advanced swept gain or voltage variable gain amplifier proposed by Mussman et al. [4] is illustrated in Fig. III-7 along with key performance data. The sensitivity of this circuit at 3.5 MHz is 6µV for a 1.0 MHz bandwidth. The gain is variable from 0 to 80 dB with excellent repeatability, an important feature in array systems and for quantitative analysis in general. As the gain control voltage (VGAIN) rises from −3.3V to −0.45V the first (of the four) stages undergoes a transition from 0 to 20 dB of gain, then the second stage, etc. This assures a minimum noise level for any value of gain.

A circuit diagram of the voltage variable gain block is illustrated in Fig. III-8. The common emitter differential input pair provides low noise and high gain bandwidth product. The dual emitter follower output buffers permit

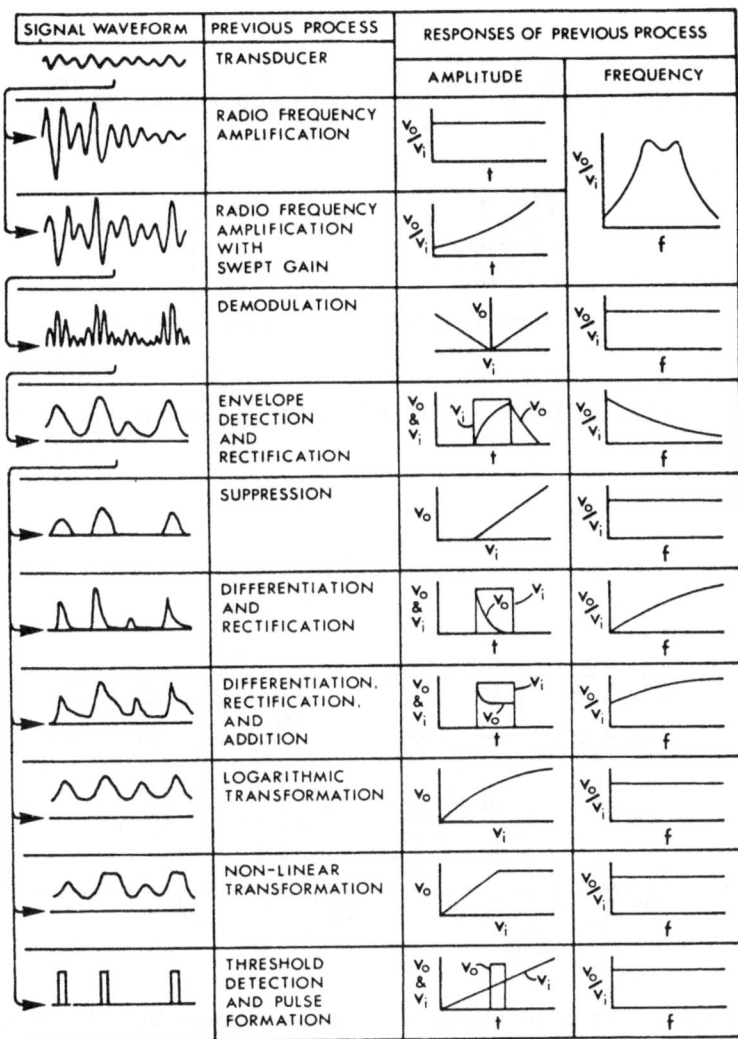

Fig. III-5. The various processing options in the pulse-echo
receiver. In each process, v_i = input voltage;
v_o = output voltage; t = time; f = frequency. See
text for detailed explanation.

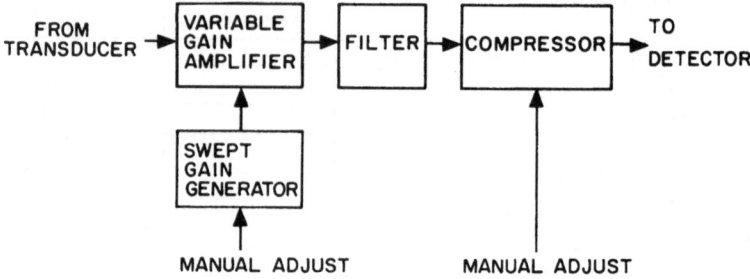

Fig. III-6.   RF amplifier block diagram

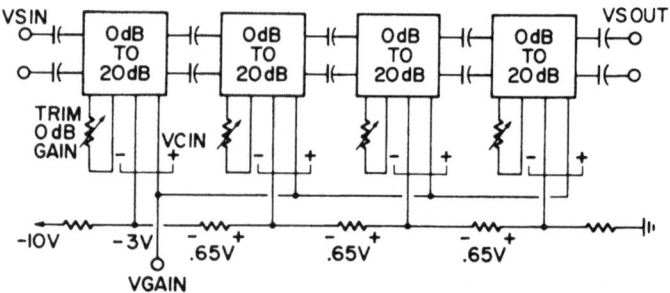

Fig. III-7. Voltage variable gain amplifier; VGAIN $\leqslant -3.3$V,
gain is 0 dB with maximum VSIN = 1.4V peak;
VGAIN $\geqslant -.45$V, gain is 80 dB with equivalent
input noise 6 nV rms$/\sqrt{\text{Hz}}$. Response is flat to
6 MHz. (after Mussman)

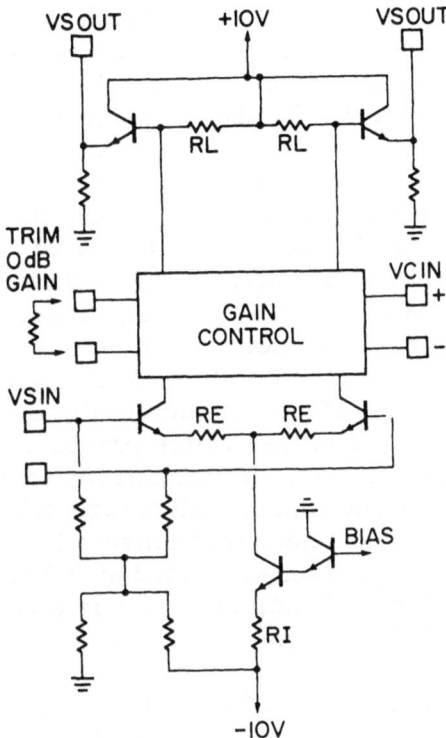

Fig. III-8. Voltage variable gain block 0 dB to 20 dB;
maximum VSIN = 1.4V peak; equivalent
input noise 6 nV rms/$\sqrt{\text{Hz}}$ at 20 dB gain,
150 nV rms/$\sqrt{\text{Hz}}$ at 0 dB gain (after Mussman)

cascading a variable number of these stages in a modular
fashion, as Fig. III-7 illustrates. The gain block itself, as
illustrated in Fig. III-8, operates in a linear fashion for peak
input voltages (VSIN) as large as 1.4V. If a matched pair of
junction field-effect transistors is substituted for the bipolar
input pair shown in Fig. III-8, the sensitivity can be improved
from 6μV to 2μV for a 1.0 MHz bandwidth.

The heart of the variable gain block is the gain control circuit illustrated in Fig. III-9 [4]. This control circuit is based upon a four quadrant multiplier. For a gain control voltage (VCIN) of $-0.3V$, Q_4 (and not Q_3) conducts most of the current from Q_2. This causes the currents in Q_1 and Q_2 to be nearly equal, as determined by the ratio of R1 to R2. Consequently, the currents in Q_6, Q_7, Q_8 and Q_9 are nearly in balance, causing ISIN$\gg$ISOUT or a gain of -20 dB for the gain control circuit. Alternatively, if VCIN $= +0.6V$, most of the current from Q_2 is diverted from Q_4 to Q_3. Thus, the current in Q_5 and hence in both Q_7 and Q_8 is considerably larger than the current in Q_4 and hence in Q_6 and Q_7. The result is that ISIN$\simeq$ISOUT is conducted almost entirely by Q_7 and Q_8 and not by Q_6 and Q_9, which provides a gain of 0 dB for the gain control circuit. In comparison to many previous approaches [1], the outstanding advantage offered by this circuit is a predictable linear relationship between gain (in dB) and control voltage over an extended (80 dB) range [4]. As future discussion will indicate, this is a consequence of its custom monolithic design.

A block diagram of an advanced logarithmic gain compression circuit based on the principle of successive limiting is illustrated in Fig. III-10 [4]. This five-stage circuit utilizes a combined summing rather than a separate summing technique to achieve a flexible modular design. The double amplifier modular building block is shown in Fig. III-11. If the TTLIN input is high (+3V), the interior differential amplifier pair is inactive and the gain provided by the exterior pair, Q_1 and Q_2, is 0 dB. In this state the maximum signal input is VSIN $= 2.75V$ for a linear response. When the TTLIN input is low (+0.5V), both the interior and exterior differential pairs are active, the interior pair providing a gain of approximately 10 dB for low level inputs (VSIN) which drops off to 0 dB for inputs VSIN$\geqslant 0.16V$. If the TTLIN inputs of all five stages are low, a 60 dB signal input range (from 0.5V to 0.5 mV) is compressed into a 20 dB output range (from 0.5V to 0.05V). Intermediate values of compression can be achieved by variation of the five TTLIN inputs which are manually programmable.

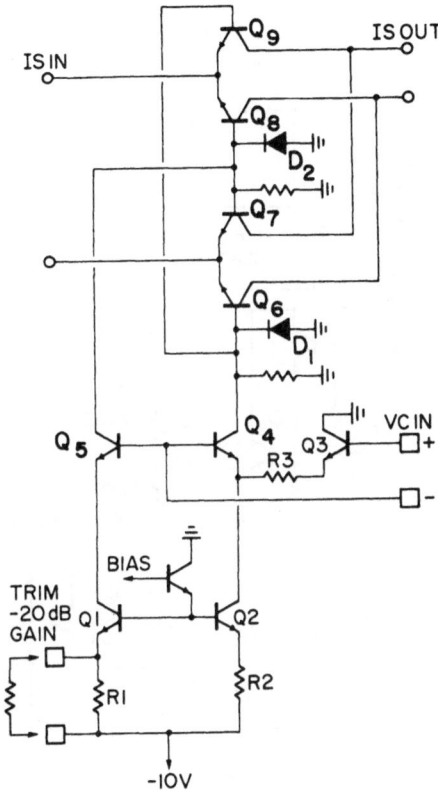

Fig. III-9.   Gain control circuit: VCIN≤-.3V, R1 and R2 set
current ratio, so gain = -20 dB; VCIN≥0.6V, Q3
diverts A2's current, gain is 0 dB (after Mussman)

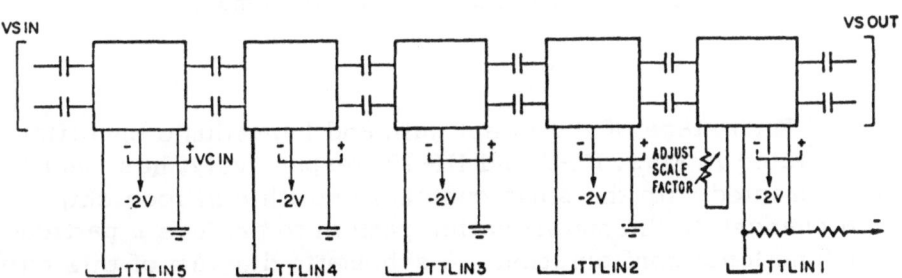

Fig. III-10.   Compressor block diagram (after Mussman)

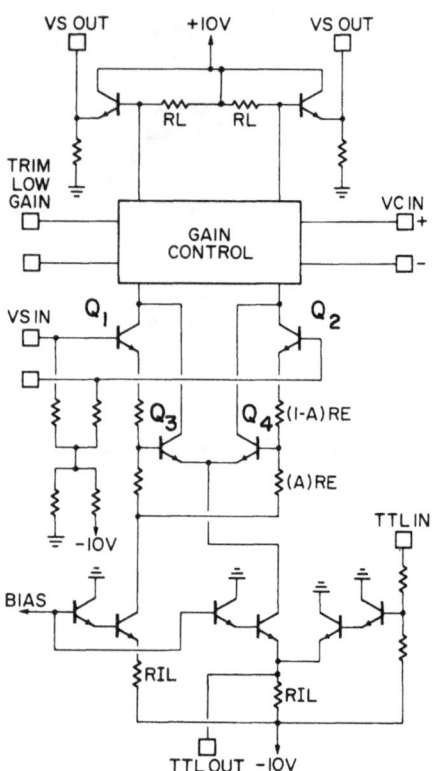

Fig. III-11. Controllable compressor block; TTLIN = 3V,
         gain = 0 dB with maximum VSIN = 2.75V peak;
         TTLIN = 0.5V, gain = 10 dB at low levels,
         dropping to 0 dB above a breakpoint of
         VSIN = 0.16V peak   (after Mussman)

   Each stage of the swept gain and logarithmic amplifiers
illustrated in Figs. III-7 and III-10, respectively, has been
implemented with the same custom monolithic silicon chip
via changes in the metallization pattern to achieve a particular functional configuration. A schematic diagram of this chip
is given in Fig. III-12 [4]. The advanced performance which

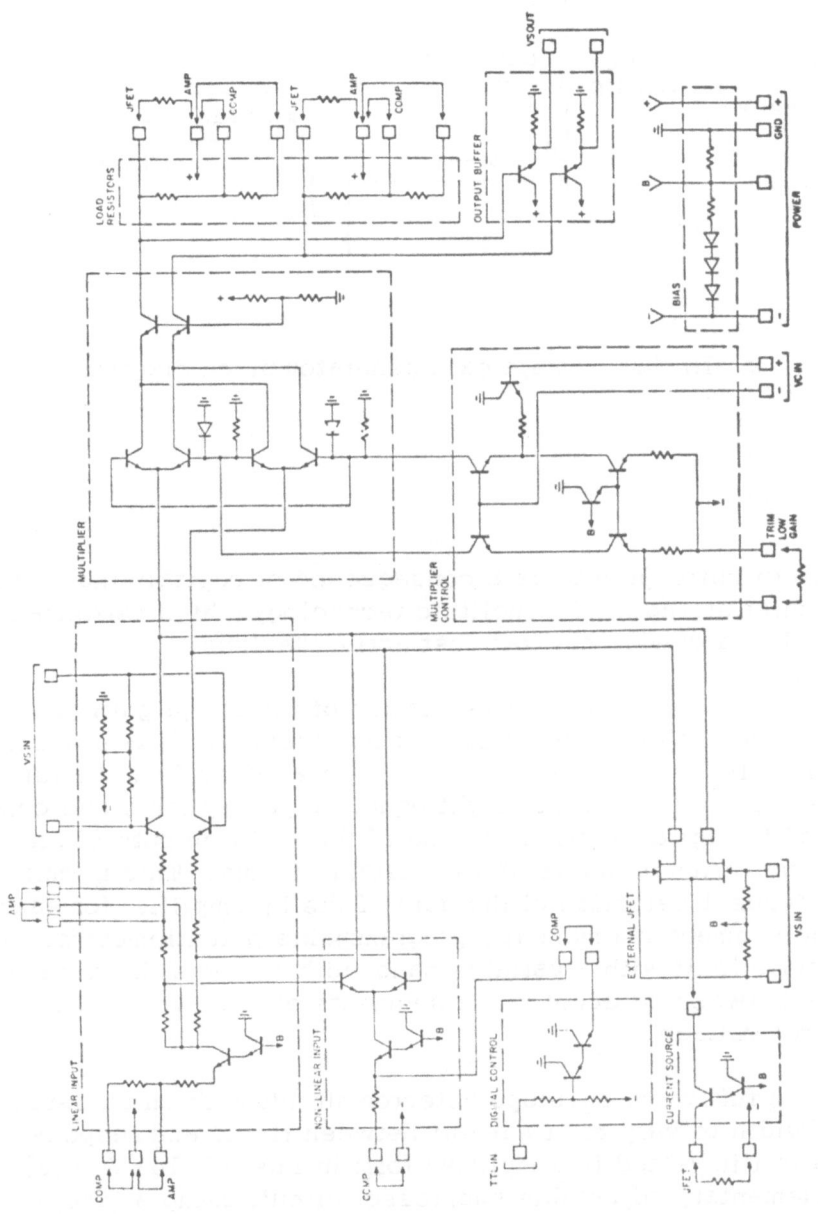

Fig. III-12. Monolithic chip layout (after Mussman)

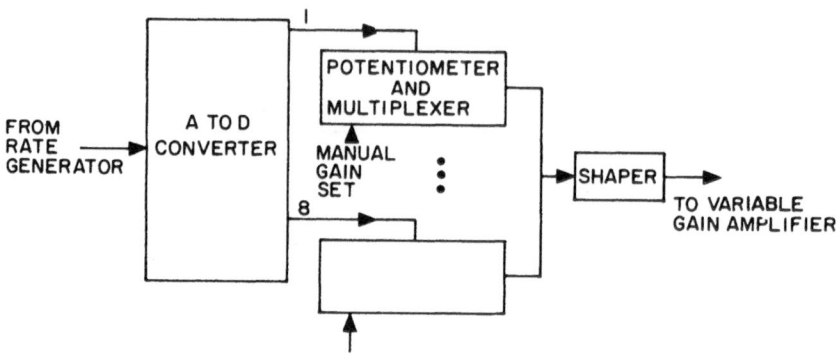

Fig. III-13.   Swept gain generator block diagram

these circuits provide is a consequence of capitalizing on the
the chief strength of monolithic technology--high performance,
closely matched, very low cost active devices.

A more detailed representation of the swept gain generator block of Fig. III-6 proposed by Mussman [5] is presented
in Fig. III-13. In this design a ramp waveform from the rate
generator is divided into eight equal range segments and converted to digital form. Only one of the eight outputs of the
A/D converter is active at any instant of time. This permits
independent selection of the gain of the RF amplifier for each
range segment via manually programmable potentiometers. A
shaper circuit with a variable bandwidth smooths the transi-
tions between adjacent range segments of the swept gain
control voltage.

A full wave envelope detector circuit with an adjustable
bandwidth to vary the trade-off between ripple and response
time is illustrated in simplified form in Fig. III-14a & b [5].
An elementary adjustable suppressor circuit using a reverse
bias diode is shown in Fig. III-14c.

(a)

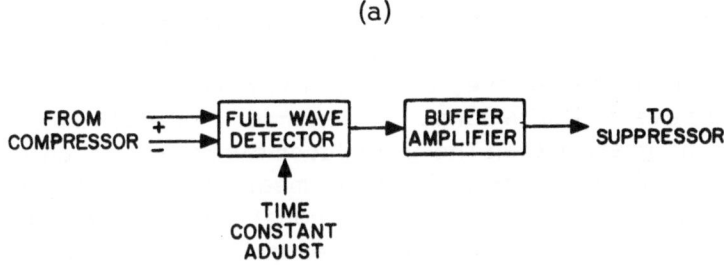

(b)

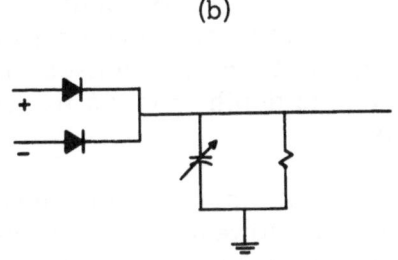

(c)

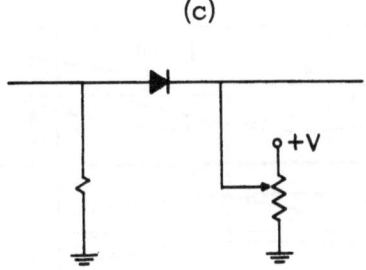

Fig. III-14.   (a) Detector block diagram
              (b) Detector schematic diagram
              (c) Suppressor schematic diagram

2) Acoustic Focus

In terms of the performance requirements summarized in Fig. II-1, the single transducer, manually scanned unfocused system described above, although extremely useful, is subject to several limitations. For example, lateral resolution can be no greater than the transducer diameter, which typically is 1.0 cm or more. A simple means of improving lateral resolution through the use of an acoustic lens bonded to the face of the transducer is illustrated in Fig. III-15 [1]. Here A is the radius of curvature of the lens and r the transducer radius. However, it is clear from this illustration that the lateral resolution of these focused transducers degrades substantially outside a limited zone of focus and may become poorer than that of an unfocused transducer of equal radius. The physical properties of acoustic lenses impose a definite lateral resolution versus depth of focus trade-off on focused transducer systems.

The field of view of a single transducer system is basically one-dimensional. However, through the use of a mechanical arm equipped with position and orientation sensing potentiometers, the field of view can be extended to two dimensions as suggested by Fig. III-3.

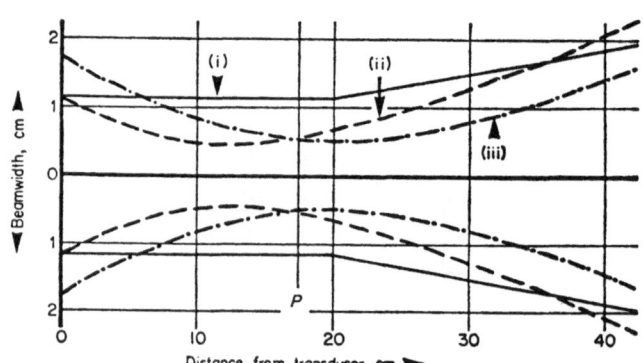

Fig. III-15. Beamwidths of focused transducers.
(i) r = 1.15 cm, A = ∞; (ii) r = 1.15 cm,
A = 17.5 cm; (iii) r = 1.75 cm, A = 25 cm
(after Wells)

The useful range of single transducer systems, either unfocused or focused, is largely determined by the required lateral resolution. The collimated beam of an unfocused transducer extends to a distance X_{max} r^2/λ where r is the transducer radius and λ the wavelength of the ultrasonic excitation [1].

For a one-dimensional field of view, the frame rate of a single transducer system is ample for virtually all real-time medical imaging applications. In this case, the frame rate is equal to the transmit pulse repetition rate (which may be 2500 pulses/second for a 30 cm range). However, if manual scanning is retained when a two-dimensional field is imaged, the frame time may extend beyond one minute. Thus accurate cardiac imaging may be impossible and fetal imaging perhaps marginal.

B. Mechanical Scan

Several forms of motor-driven mechanical scanning of a single transducer have been used to achieve real-time two-dimensional imaging of the heart [6,7], the eye [8] and the abdomen [9]. One of these motor driven scanners as proposed by Griffith and Henry [6] is illustrated in Fig. III-16.

In general, mechanical scanning of a single transducer represents a simple and low cost method of extending the performance of a manually scanned system to achieve two-dimensional fields of view in real time. However, such systems may be limited in some respects in comparison to the possibilities of electronically scanned and focused systems. These limitations include: a) a much more restricted field of view or zone of high resolution, b) little potential for extending the field of view to a three-dimensional volume to achieve high resolution, real-time volumetric imaging, c) greater difficulty in coupling the transducer to the patient because of its oscillatory motion, and d) less flexibility in other aspects of performance due to the cumbersome qualities of motor-driven mechanical components compared with electronics.

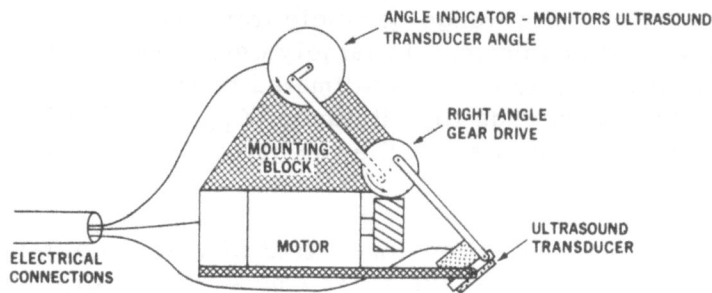

Fig. III-16. Diagrammatic representation of the hand-held
scanner. A tachometer-feedback regulated DC
motor drives a crank and lever system which
angles the ultrasound transducer in a recipro-
cating manner through a 30-degree sector. The
angle indicator is driven simultaneously by an
identical mechanical linkage and generates an
electrical signal proportional to the transducer
angle. (after Griffith)

IV.   LINEAR ARRAY SYSTEMS

The primary goal of this section is to review the
salient features of ultrasonic imaging systems which are
based upon linear transducer arrays of various types. Electronic delay lines and multiplex switches assume major
importance in connection with linear arrays, whereas they
are of no consequence in single transducer systems.

A. Manual Scan

As noted in the previous section of the discussion
related to Fig. III-15, one of the chief limitations of a
focused single transducer is its restricted depth of focus.
This limitation can be overcome to a large extent if an electronic focusing technique is used in place of an acoustic
lens. In essence, electronic focusing entails the use of an
array of transducers, and several approaches have been
reported which might be classified as "linear array manually

scanned systems" [1,2,3]. Perhaps the most advanced of these approaches, proposed by Burckhardt et al. [3] exploits the fact that a thin annular transducer is effectively in focus at all points along its axis. This yields high lateral resolution over a large depth of field, as suggested by Fig. IV-1a, because on every point of the axis the contributions of all elements of the annulus add in phase, whereas off axis they do not. To overcome the large sidelobes that accompany a focused annular transducer, a segmented annulus is used as illustrated by Fig. IV-1b. Two successive pulses are used for sidelobe reduction, and between them the transmit/receive configuration of the eight annular segments is rotated by 45°, using electronic multiplex switches, as illustrated in Fig. IV-1b & c. The echo signals from the two pulses are added as indicated by the system block diagram of Fig. IV-2.

This block diagram reveals somewhat greater electronic complexity than is apparent in the single transducer system of Fig. III-1. Specific additions include: a) a transmit power amplifier and a receiver preamplifier associated with each of the eight annular segments, 2) two sets of multiple switches needed to rotate the transmit/receive pattern, 3) two multipliers, A/D converters, digital shift register delay lines and adders used to sum echo signals from two successive transmit pulses, and 4) various other circuits.

The annular array system diagrammed in Fig. IV-2 apparently provides substantial improvement in depth of focus for a given lateral resolution compared with focused single transducer systems. It offers high resolution in all three spatial dimensions as well as a two-dimensional field of view if the transducer is incorporated in a scanning arm with potentiometers to indicate position and orientation. A full three-dimensional field of view is unavailable. If the annular transducer array is incorporated in a motor-driven mechanical sector scanner, real-time imaging is quite feasible, although the frame time is doubled by the requirement of two scans per line. However, some form of electronic sector scanning may be a preferable technique for an electronically focused system of this type.

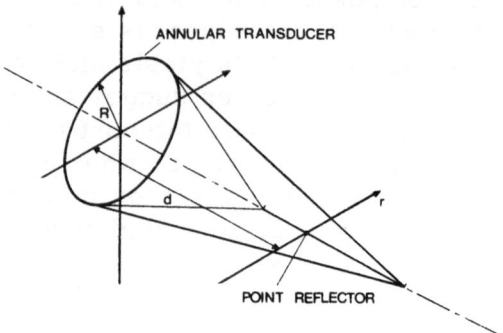

Fig. IV-1a. Perspective view of the annular transducer
             (after Burckhardt)

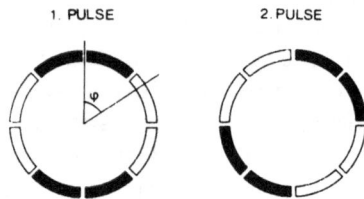

Fig. IV-1b. Transmitting and receiving segments for the first
             and second pulse. Black: transmitting. White:
             receiving. (after Burckhardt)

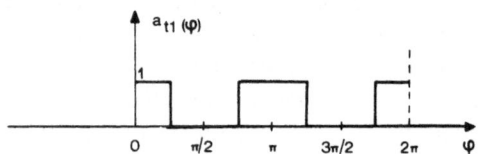

F ig. IV-1c. Transmitted amplitude $a_{t1}(\varphi)$ for the first pulse
              as a function of angle φ on the ring   (after
              Burckhardt)

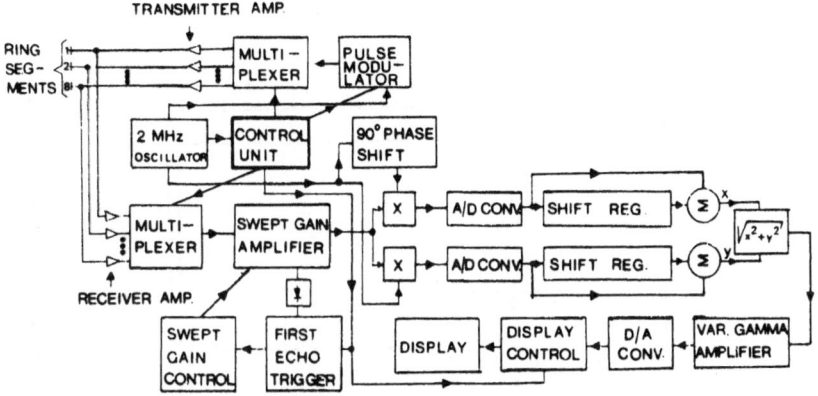

Fig. IV-2.   Block diagram of the experimental system
              (after Burckhardt)

B. Mechanical Scan

1) Acoustic Focus

A block diagram of a linear array, mechanically scanned,
acoustically focused imaging system proposed by Green et al.[4]
is illustrated in Fig. IV-3. Unlike previous systems considered
here, this system does not provide either A-scan or B-scan
operation. It is a C-scan system, i.e. the image plane is
orthogonal rather than parallel to the axis of the transmitted
sound. The system uses a linear array of 192 separate transducers which operate in the receive mode only. Counter-rotating
acoustic prisms refract sound waves from the object to provide,
in effect, a linear mechanical scan of the transducer array over
the image plane. A pair of acoustic lenses with a 1.0 cm zone
of focus serve to form the image in the plane of the transducer
array. As suggested by Fig. IV-3, this system may operate in
either a transmission or reflection mode.

In terms of electronics there is no need for swept gain
amplifiers to compensate for tissue absorption, but 192 closely
matched high gain linear/logarithmic amplifier channels are

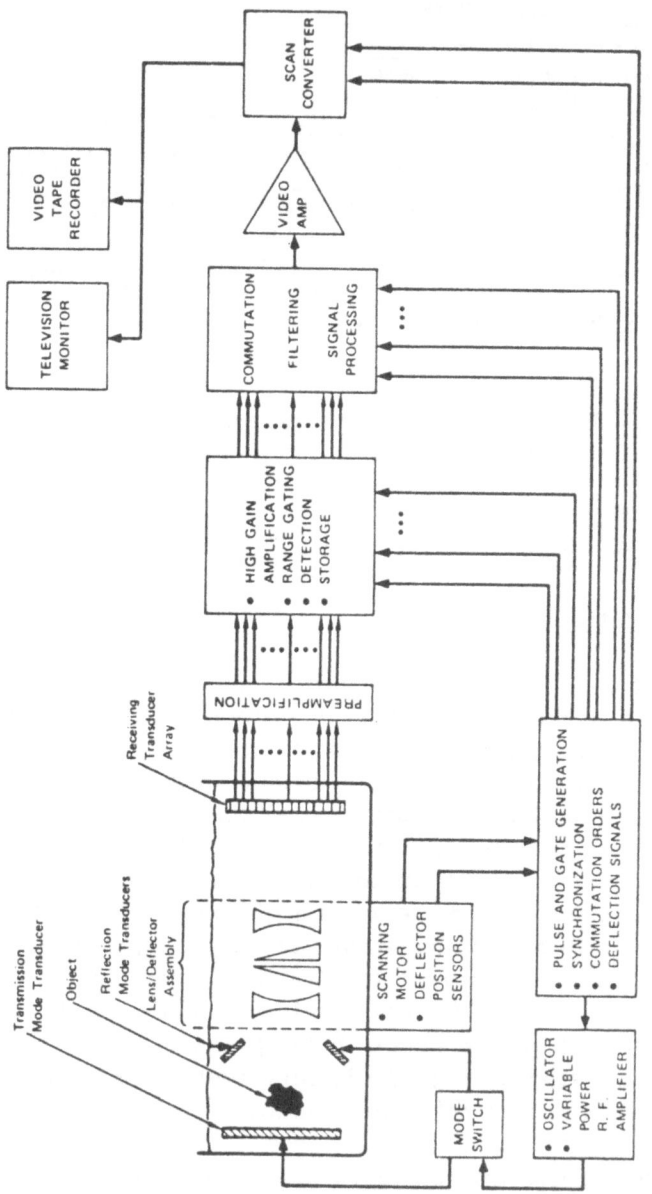

Fig. IV-3.　Block diagram of the SRI ultrasonic camera system (after Green)

used for the receiving array, since multiplexing (commutation) is a post-detection process in this system. A single transmitter, no delay lines and an apparently substantial amount of scanning electronics are needed.

In the reflection mode of operation high axial resolution is difficult to achieve due to the physical separation of the transmit and receive transducers and their relative motion. In the transmission mode axial resolution is an undetermined parameter. Lateral resolution, as determined by a large aperture acoustic lens system, may be excellent. However, "out-of-focus" signals in the transmission mode and multiple reflections resulting from "flooding" the entire object field with transmitter output in the reflection mode, may produce blurring. The field of view of the system is a relatively large orthogonal plane. A three-dimensional volumetric field is not available. Optimum use of body apertures is not possible in the transmission mode for obvious reasons. To a relatively minor degree this is also true in the reflection mode, since the transmit and receive transducers are physically separate. Range requirements may or may not be more difficult to meet than in A- or B-scan systems, depending on the nature of the transmission path; in the reflection mode losses and undesired reflections due to the multiple acoustic components may be more troublesome than those in direct contact systems. Real-time operation is feasible, although turbulent effects caused by the rotating acoustic prisms must be avoided.

2) Electronic Focus

In essence, the system described by Fig. IV-2, if provided with a motor-driven mechanical sector scanner, would fit this sub-classification. No such system has yet been reported, although the approach appears promising.

3) Holographic

The system proposed by Fenner et al. [5] shown in block diagram form in Fig. IV-4a & b can be classified as a linear array, mechanically scanned holographic system. Image

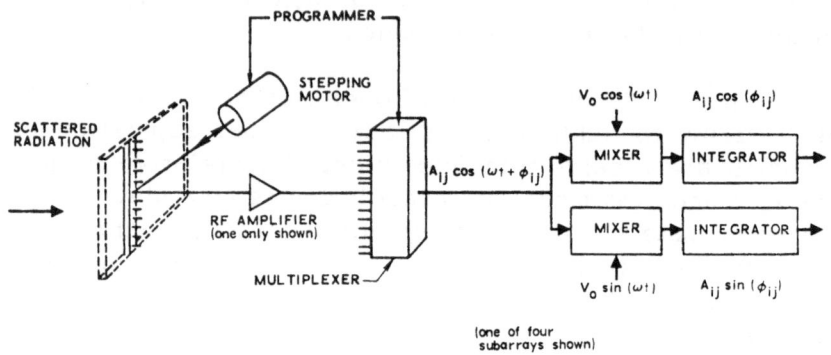

Fig. IV-4a. Mechanically scanned linear array (after Fenner)

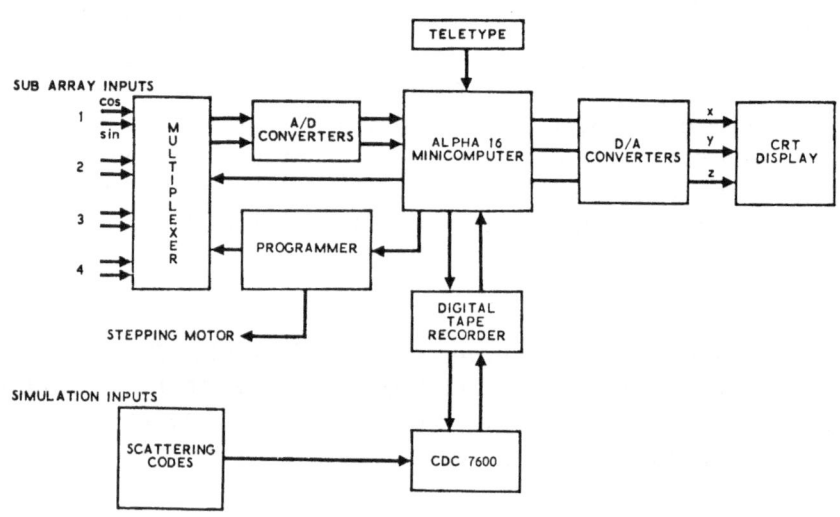

Fig. IV-4b. System block diagram (after Fenner)

reconstruction is achieved through digital storage and subsequent processing of instantaneous amplitude and phase information collected from each element of an entire two-dimensional image plane. The front end electronics of this system includes sensitive receivers, quadrature mixers, multiplexers and A/D converters. Experimental data describing the performance of the system is not available. A potential for high resolution, large three-dimensional fields of view and adequate range exists; real-time operation would require extremely high speed digital processing. Some fundamental problems of coherent systems, including speckle noise and out-of-focus artifacts, continue to be important considerations.

C. Electronic Scan

1) Rectilinear Scan

A block diagram of a linear array, rectilinear electronic-scan unfocused pulse-echo imaging system proposed by Bom et al. [6] is illustrated in Fig. IV-5. The transducer and display arrangement is sketched in Fig. IV-6. The concept of this multiple B-scan system is straightforward. Through an array of time division multiplex switches, an entire linear array of transducers shares a common set of receiver, transmitter and control electronics, with the exception that a preamplifier is used in conjunction with each transducer element. Closely matched preamplifiers and high voltage multiplex switches for the transmitter driver present the most severe demands for high performance electronics.

Although this system is capable of high axial resolution, lateral resolution degrades significantly with range as indicated in Fig. IV-7. The field of view is a plane which is restricted by the boundaries of the transducer array. Compared with a sector scanner this is an advantage at short range and a disadvantage at long range. A volumetric field of view is not available. Range may be reduced relative to larger diameter single transducer systems due to reductions in power density accompanying beam spreading. Real-time cardiac imaging is easily achieved.

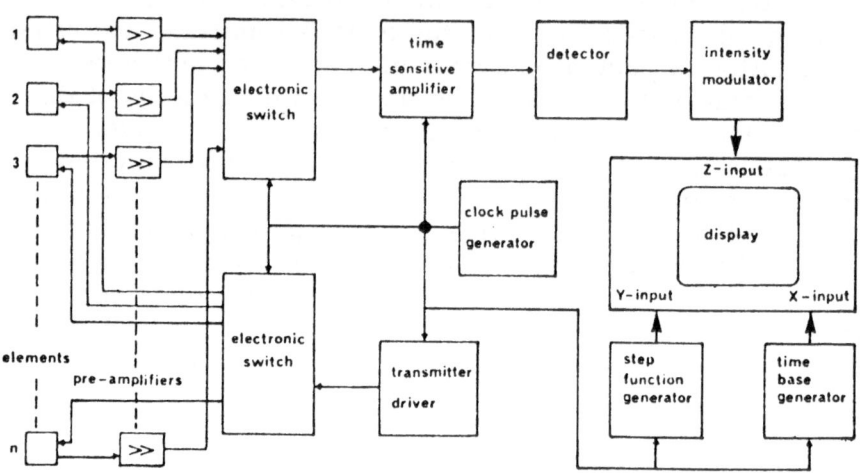

Fig. IV-5. Block diagram of an n-element system (after Bom)

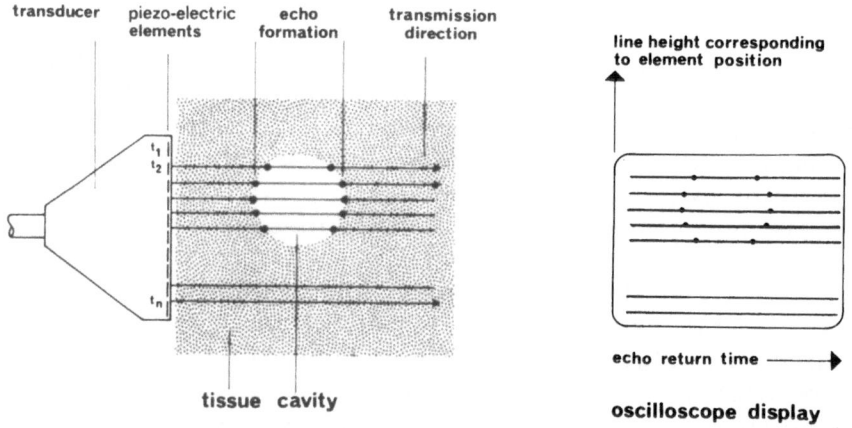

Fig. IV-6. Schematic drawing of a multi-element transducer
            and corresponding display (after Bom)

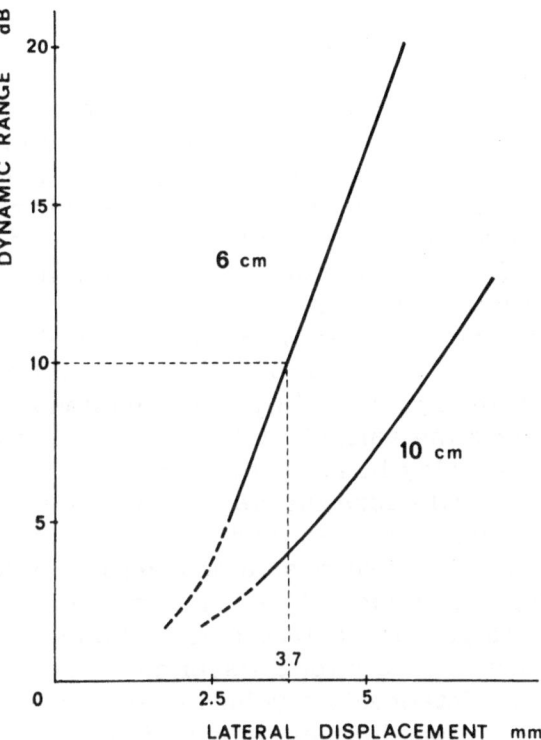

Fig. IV-7. Lateral resolution of a single 4 mm side element from the Multiscan transducer with the intermediate frequency (3.5 MHz) and measured at 6 and 10 cm distance. When the echo from a point reflector along the main axis of the element is just visible at 0 dB, it will no longer be visible if moved laterally over any distance. If the echo strength is increased 10 dB, a reflector at 6 cm depth may be moved 3.7 mm in either direction and will still be visible (dotted line), while a reflector at 10 cm depth can be moved 6 mm laterally and still be visible. The limits of lateral resolution for a range of intensities can be determined in this way. (after Bom)

2) Sector Scan

a) Unfocused

A linear array, sector scan, unfocused imaging system proposed by Somer [7] is illustrated in Fig. IV-8. Note that each transducer element requires a separate transmitter. By energizing the transmitter array elements in the proper time sequence--i.e. with a time delay of $d \frac{\sin \theta}{c}$ (where c is the sonic velocity and d is the inter-element distance) between adjacent element pulses--a plane wave can be transmitted at an angle θ to the normal of the array as shown in Fig. IV-9. Reception of the echoes from this wave is illustrated in Fig. IV-10 [8]. To understand the principle, one may assume the variable delay time $\tau = 0$ for all sections of the variable delay line. In this instance the fixed delays t_1 through t_{20} will cause the transducer signals from the incoming wavefront at an angle θ to sum in phase at the amplifier input. Now suppose a non-zero time delay τ given by $20\tau = 2t_1$ is provided by each section of variable delay line. Here the pattern of fixed plus variable time delays causes the array signals for a wavefront at an angle $-\theta$ to sum in phase. Intermediate values of the beam steering angle can be achieved for alternate values of τ.

Apparently, the fixed delays used in this system are achieved using distributed coaxial delay lines, while the variable delay times are produced by a gated capacitance store delay network as illustrated in Fig. IV-11 [8]. The pairs of switches connected by dashed lines operate simultaneously in a time sequence described by the waveforms of Fig. IV-12 to achieve a delay (t_d) bandwidth (B) product given by $Bt_d = (N-1)/2$. One of the $6 \times 20 = 120$ variable delay line stages needed for this system is diagrammed in Fig. IV-13. Clearly the transmitters and receiver delay lines used in this system represent a substantial increase in component count compared with single transducer systems.

The axial and lateral resolution of this system are essentially equivalent to those of an unfocused single

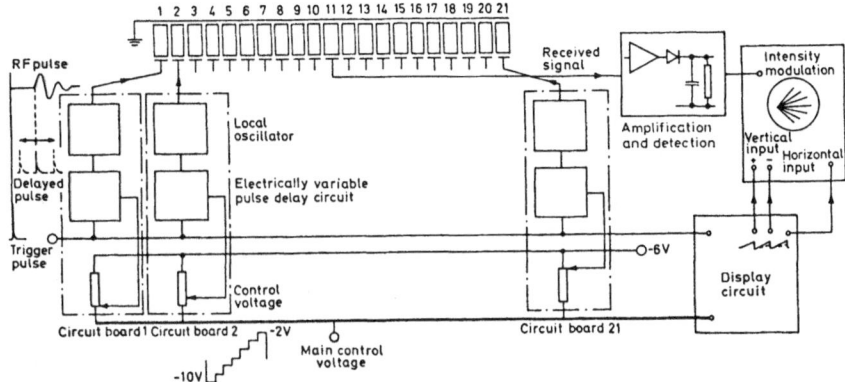

Fig. IV-8. Block diagram of sector scan equipment (after Somer)

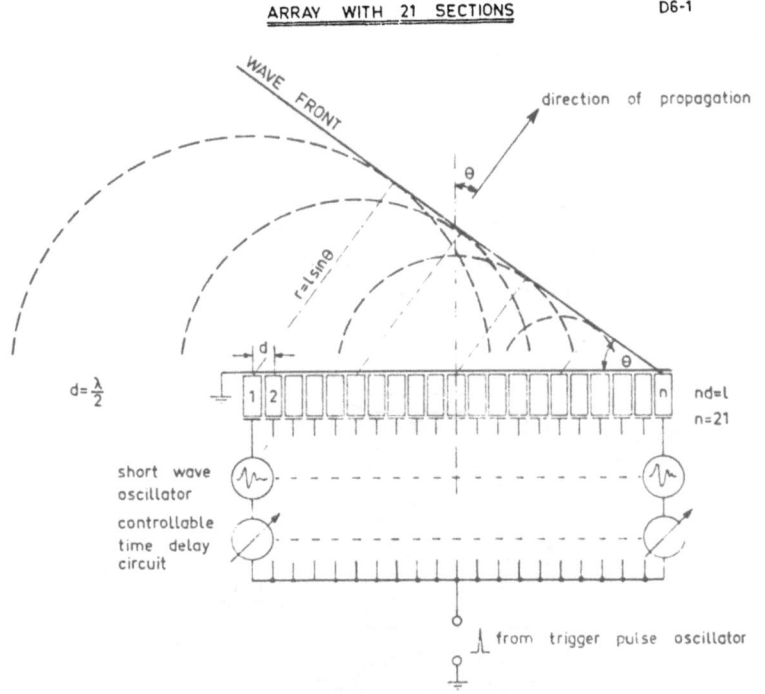

Fig. IV-9. Principle of forming a transmitted beam with an arbitrary direction, by the use of an array of 21 elements with local oscillators and pulse-delaying circuits (after Somer)

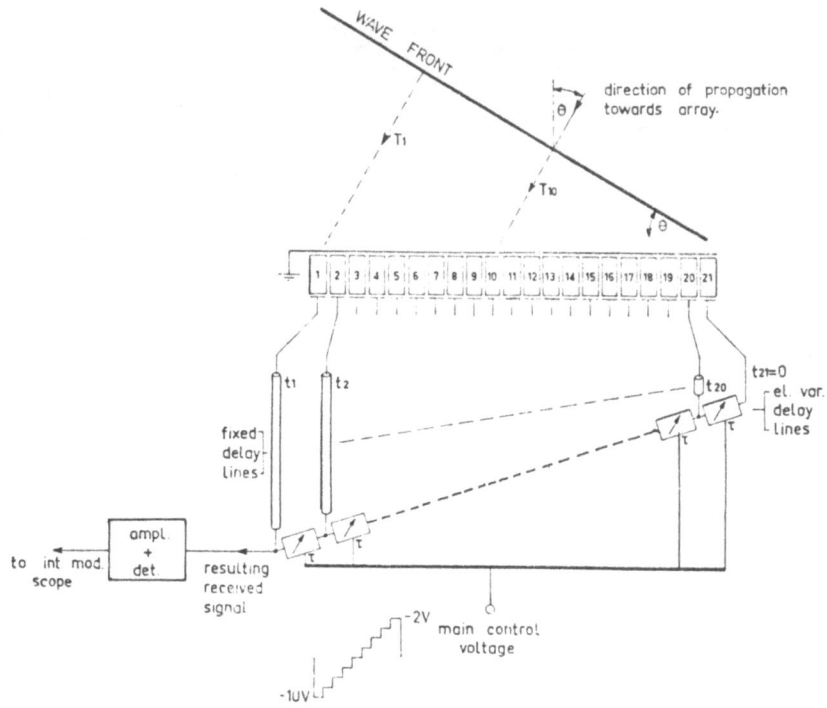

Fig. IV-10. Principle of forming a resolved beam with an
arbitrary direction using a system of fixed and
electrically variable delay lines (after Somer)

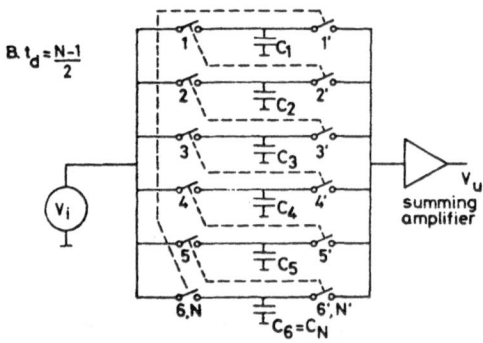

Fig. IV-11. Gated-capacitance-store delay network (after
Somer)

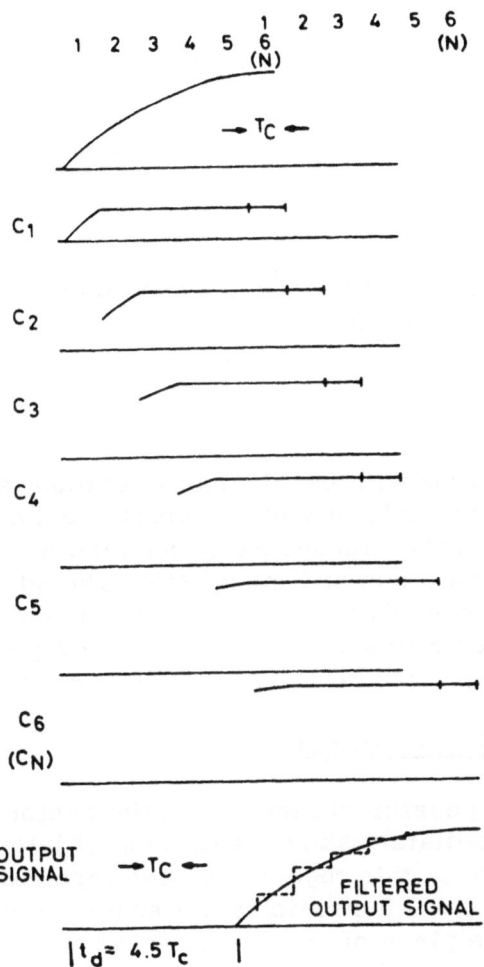

Fig. IV-12. Voltages in the delay network (after Somer)

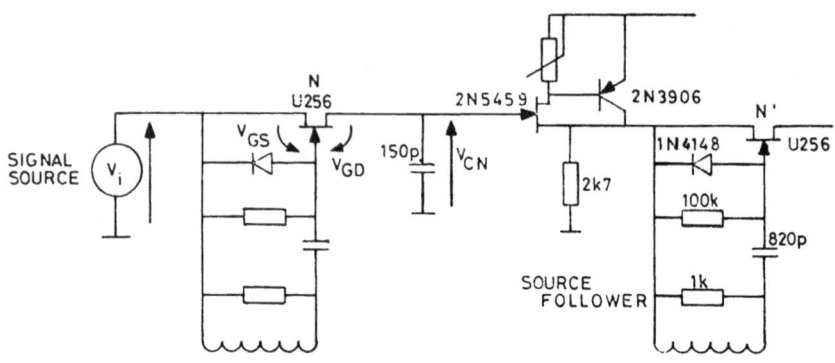

Fig. IV-13. Circuit schematic for one stage of the delay line
(after Somer)

transducer system with equal transducer dimensions. The
two-dimensional field of view is largely equivalent to that of
an unfocused single transducer mechanical sector scan system.
Likewise for the range and frame rate. The advantages which
are offered compared with an unfocused single transducer system include easier coupling to the body and greater flexibility
in adjusting field of view, range and frame rate.

b) Cylindrical Focus

      A possible deficiency of the sector scan system
described immediately above is its lack of lateral resolution.
An improvement in this regard has been reported by Thurstone[9],
who in addition to electronic sector scanning uses electronic
focusing in the plane of the scan. This improves lateral resolution in the plane of the scan but results in no improvement
in lateral resolution orthogonal to the scan plane.

      A block diagram of this system is illustrated in
Fig. IV-14 [9]. Note that each of the 16 transducer elements
has a separate preamplifier, logarithmic compressor, variable

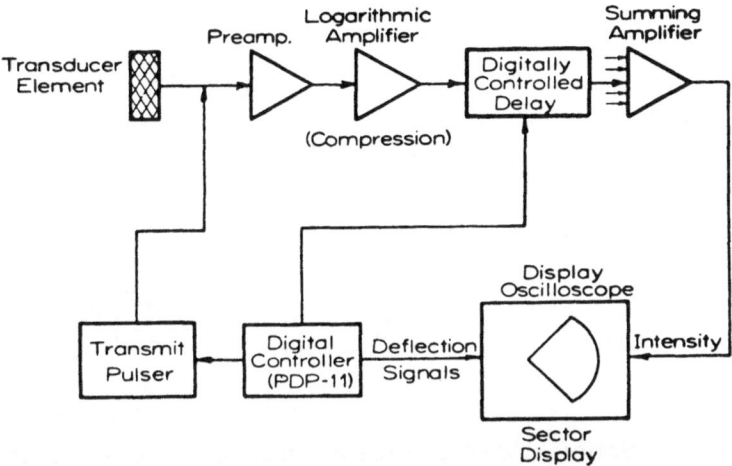

Fig. IV-14. Block diagram of one channel of the sector scan imaging system (after Thurstone)

delay line and transmitter associated with it. Beyond the use of electronic focusing of received signals _per se_, some additional lateral resolution in the plane of the scan is achieved by multiplying (i.e. adding the logs) the signals of the individual elements and by using an aspheric curve for the transmit beam [9,10]. This process, however, makes the system nonlinear so that undesired artifacts can be generated. The delay lines apparently are lumped constant LC networks with transistor taps controlled by a dedicated PDP-11 mini-computer which also controls the aspheric transmit beam.

The flexibility of this system is remarkable. Additional features which would be useful in some applications are improved lateral resolution orthogonal to the scan plane and the capability to provide a three-dimensional volumetric field of view.

3) Orthogonal Scan

A distinctive approach to ultrasonic imaging reported by Kino et al. [11,12] is illustrated by the diagram of Fig. IV-15.

TRANSMITTER

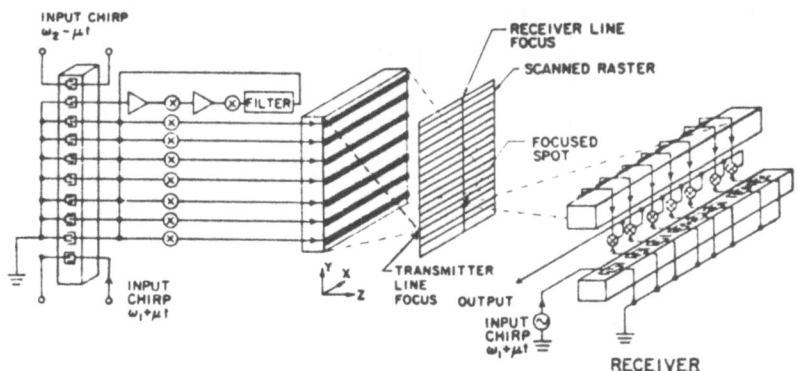

Fig. IV-15. Schematic-pictorial diagram of an electronically
scanned, electronically focused acoustic imaging
device (after Kino)

In this approach a nominally 50 MHz chirp signal is fed into a
surface acoustic wave (SAW) delay line which is uniformly
taped to provide a linear array of output signals which are
mixed with 2.25 MHz signals from a linear array of piezoelectric transducers to form a single output signal at the sum or
difference frequency. This nominally 50 MHz signal then
undergoes further processing toward a CRT display. The key
feature of the receiver is that the square law variation of phase
of the signals arriving at the transducer array from a particular
vertical line of the object must be cancelled by an equal but
opposite square law phase variation in the signals from the
SAW delay line, thus producing no phase variation in the output signal and effectively focusing the receiver array on the
vertical object line. The transmitter SAW delay line is excited
from opposite ends (using nominally 200, 250 and 300 MHz
oscillators) to produce two different chirp signals which are
combined acoustically and mixed down electrically to 2.25 MHz
to excite the transmitter array and produce a horizontal line of
focus. Scan lines in this system are essentially orthogonal to
the axis of travel of transmitted waves. In fact, one might
say that in some sense the principle of operation of the entire
system is orthogonal to that of a B-scan system.

At present, performance data defining the resolution, field of view, range and frame rate of this system in medical applications is lacking. However, the system does focus in all three spatial dimensions, although it might be anticipated that phase ambiguities of little consequence in pure time delay systems will influence both axial and lateral resolution. Since the system does not employ beam steering, its volumetric field of view is limited to the shadow of the transmit transducer. Field of view is also restricted by use of the transmission mode of operation, which may affect range. The matched UHF circuit requirements of this system precipitated by the use of SAW delay lines may be more troublesome than those of lower frequency baseband circuits.

4) Holographic

A circular array, electronic-scan, holographic imaging system which has been proposed by Vilkomerson [13] is illustrated in Fig. IV-16. This system concept calls for spatially sampling the instantaneous amplitude and phase of the acoustic signals distributed over a given aperture by using an annular array of 400 piezoelectric transducers. Computer reconstruction will be used to obtain the complete acoustic field pattern over the aperture, and from this the acoustic image of the target. This system is advanced in concept and may offer commensurate performance. To date, it has not been demonstrated in medical applications. However, it is clear that if real-time operation is desired, perhaps 400 channels of receiver electronics and extremely high speed digital processing will be necessary. As with the Burckhardt system[3], however, the circular array has limited receiver sensitivity for a given size aperture.

D. Additional Approaches

In this section an attempt was made to review a representative sample of the various types of linear array systems which have been reported. Although this was done in a rather comprehensive fashion, certain efforts such as those of Knollman et al. [14] did not receive explicit attention because of their (essential) similarity to systems which were discussed in some detail.

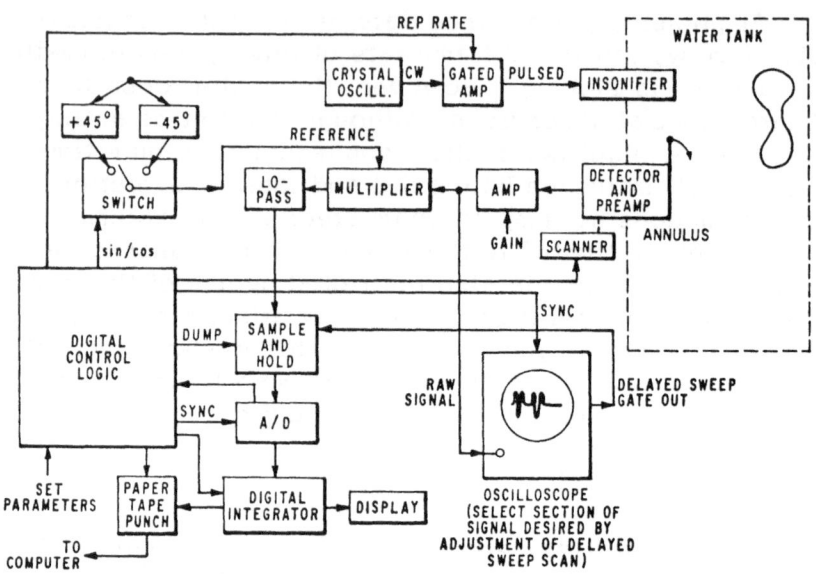

Fig. IV-16.   Electronic equipment used to measure the ampli-
              tude and phase around the thin annulus (after
              Vilkomerson)

V.   AREA ARRAY SYSTEMS

In this section ultrasonic imaging systems which
incorporate two-dimensional area arrays of transducers are
reviewed. Unusual opportunities for high spatial resolution
throughout a volumetric field of view are presented by these
systems. However, electrical interconnections to their
arrays present problems which are of relatively minor consequence in linear arrays.

A. Acoustic Focus

Approaches to ultrasonic imaging systems using area
arrays with acoustic focusing have been reported by Cook [1],
Takagi et al. [2], Harrold [3], and Maginness et al. [4]. In
this instance, the latter system provides the most appropriate
basis for discussion of the key features of area arrays.

A block diagram of the ULISYS system [4] is illustrated
in Fig. V-1. In time sequence each element in the two-
dimensional transducer array operates in a pulse-echo mode
as both a transmitter and receiver of ultrasonic energy. Two
key techniques used to achieve this are illustrated schematically in Fig. V-2. They are a set of simple x-y address lines
and a multiplex switch co-located with each transducer. The
three-layer structure which implements this scheme is illustrated in Fig. V-3. The x-address lines are deposited on the
top side of the intermediate insulating support layer and the
y-address lines on the bottom. The multiplex switches are
double-diffused metal-oxide semiconductor (DMOS) transistors [5], with high breakdown voltage (>200V), high pulse
current (>400 mA) and low on resistance (<30 ohms)--charac-
teristics which are virtually ideal for transmit/receive multiplex switches for ultrasonic imaging systems. In effect, the

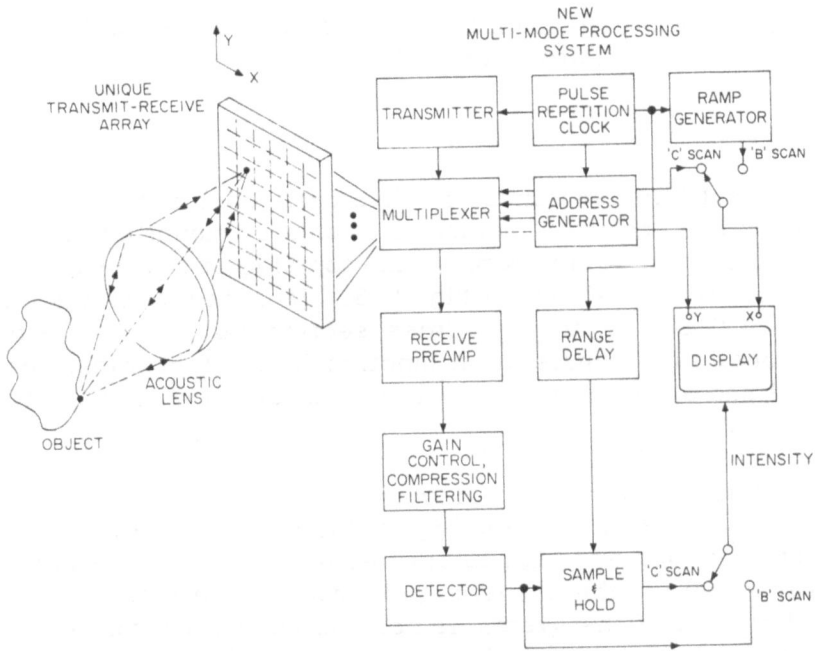

Fig. V-1. Block diagram of system (after Maginness)

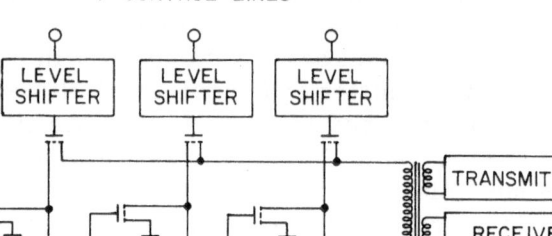

Y CONTROL LINES

Fig. V-2.   Electrical schematic of array circuitry
(after Plummer)

DMOS multiplex switches are physically co-located with
their corresponding transducer elements by mounting them in
integrated circuit chip form on the top side of the insulating
support layer as shown in Fig. V-3. Each chip contains an
array of DMOS switches. Cross-sectional diagrams of a conventional MOS transistor, a junction field effect transistor
(JFET) and a DMOS transistor are illustrated for comparison
in Fig. V-4.

     Using acoustic focusing as indicated in Fig. V-1, this
area array system provides high axial resolution via short
transmit pulses and high lateral resolution (in both lateral
dimensions) via the focusing properties of the lens. The
field of view of the system is not limited to a sector of a
plane, as is the case with a sector scanned linear array.
Instead it is a three-dimensional volumetric field with a

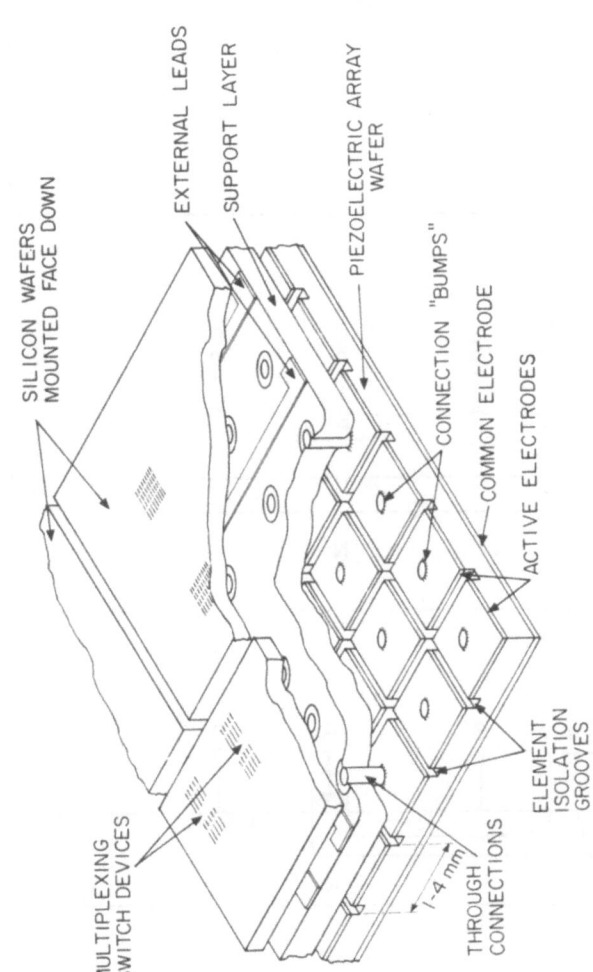

Fig. V-3. Array construction (after Maginness)

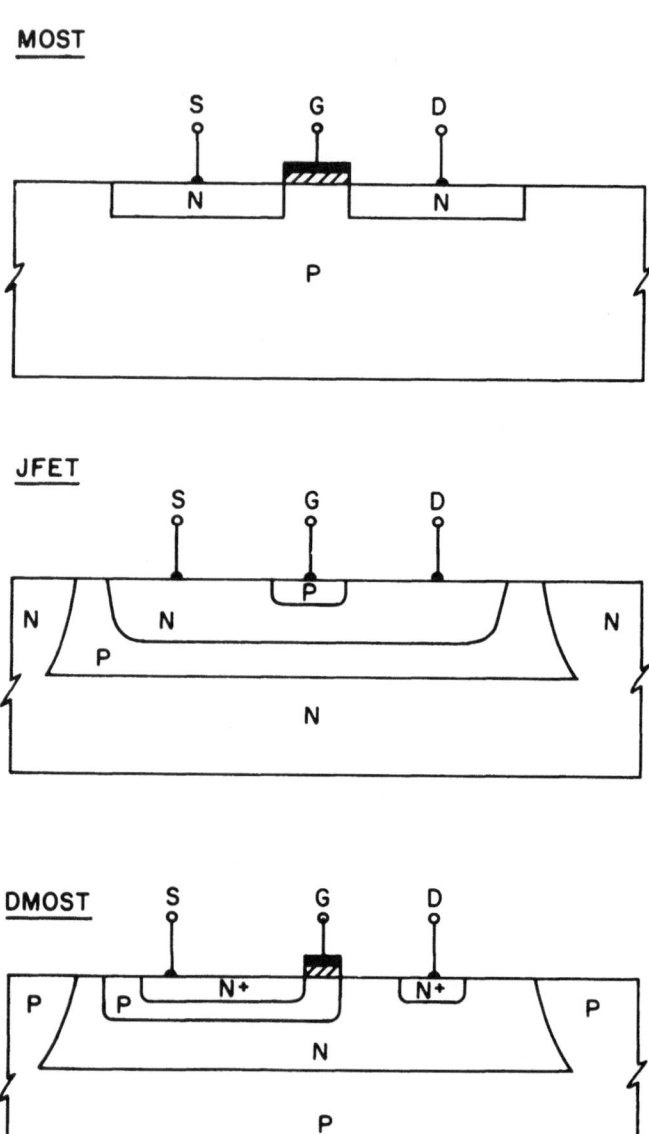

Fig. V-4. Cross-sectional views of transistor multiplex switches

pyramidal shape as illustrated in Fig. V-5 [6]. Using the
reflection mode of operation with a transmit/receive array,
this pyramidal volume provides the maximum field of view
that can be obtained for a given limited body aperture. Note
that the field can be scanned in real time on a row-by-row,
on a column-by-column or on an arbitrary basis. The range
and frame rate of the system permit deep body imaging in real
time. Accurate dynamic measurements of heart chamber
volumes are quite feasible with this system. Note that it is
capable of operation in the A-scan, B-scan or C-scan modes
and provides the high volumetric resolution and large volumetric field of view necessary for accurate measurement of
tissue parameters such as attenuation, acoustic velocity and
impedance as well as echo-scattering cross section.

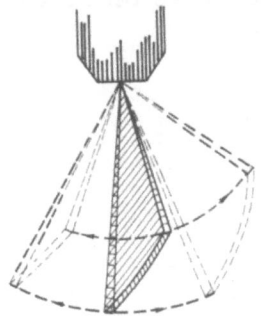   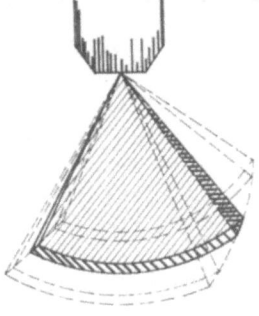

Scan by element rows   Scan by element columns

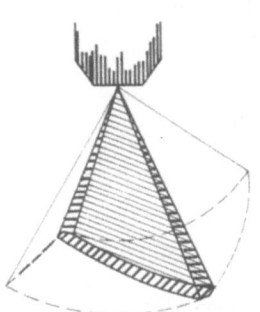   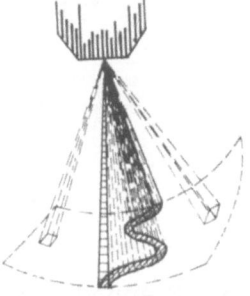

Scan by diagonal elements   Scan in an arbitrary pattern

Fig. V-5. Sector scanning in three dimensions is pyramidal
          scanning (after Beaver)

B. Electronic Focus

Electronic focusing with an area array such as that illustrated in Fig. V-1 presents a rather formidable technological problem, considering the number of transmitters, receivers, delay lines and other circuits such a system might entail. However, some promising advances toward electronically focused systems which provide performance comparable to that of the system of Fig. V-1 have been reported by Macovski and Norton[7] using a "θ" array.

The most critical electronic component, a charge coupled device (CCD) delay line necessary for the θ array system has been described by Melen et al. [8]. From the illustration in Fig. V-6 [9]), one may observe that a CCD

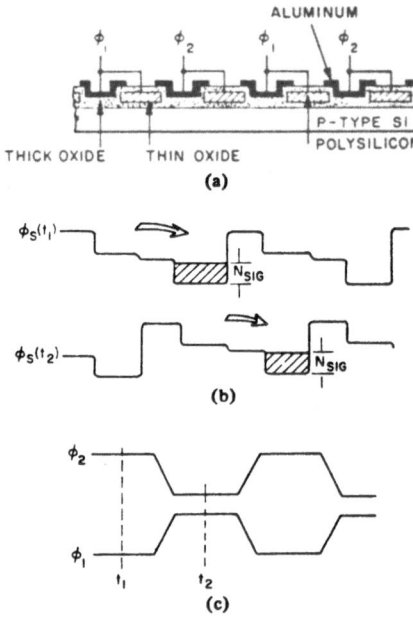

Fig. V-6. Stepped-oxide two-phase CCD. (a) Cross-sectional view of the structure (b) Surface-potential profiles showing charge transfer (c) Clocking waveforms used to drive the CCD during transfer (after Barbe)

is a series strong of metal-oxide-semiconductor (MOS) capacitors with separate upper electrode pairs and a bottom silicon
electrode common to all capacitors. Transfer of signal charge
packets of arbitrary size between adjacent capacitors is
accomplished by reducing the potential (i.e. clock voltage) of
alternate upper electrode pairs while simultaneously increasing the potential (clock voltage) on adjacent upper electrode
pairs. As these electrode potentials switch, corresponding
changes in the silicon surface potentials cause a lateral
signal charge transfer as illustrated in Fig. V-6. The salient
features of CCD electronic delay lines are the following:
a) delay time (t_d) can be varied electronically by simply
changing the clock frequency (f_c) of the CCD, b) delay time
is the product of the number of pairs of stages (N) and the
clock frequency and therefore can be extended by increasing
the number of stages, c) high charge transfer efficiency
(e.g. $\eta > 99.99\%$) between adjacent electrodes (which is required in order to use a large number of stages), d) large
transfer efficiency at clock frequencies of 10 MHz and above
(which is required to permit wideband signals to be delayed),
e) large dynamic range (e.g. ~ 60 dB), f) a time delay (t_d) -
bandwidth (B) product given by $t_d B \simeq (Nf_c)B \simeq N/2$ (using the
sampling theorem) and g) potentially low cost due to the
fact that entire arrays of CCD delay lines can be fabricated
as a monolithic silicon integrated circuit.

The use of a monolithic array of CCD delay lines with
quadratically spaced input taps which match the curvature of
spherical acoustic wavefronts is illustrated in Fig. V-7 [8].
Sweeping the clock frequency dynamically changes the focus
from target point T_1 to point T_2. The use of an array of CCD
delay lines with linearly spaced input taps for beam steering
is illustrated in Fig. V-8 [8]. Changing the frequency of
clock 1 in discrete steps changes the value of the scan
angle θ for beam steering to the right of the z-axis. A simple
input switching matrix can be used to reverse the taps (i.e.
place the longest delay line on the left) to obtain beam
steering to the left of the z-axis. Alternatively, a third array
(of linearly tapped delay lines) can be added as illustrated
in Fig. V-9 [6] to enable beam steering on either side of the

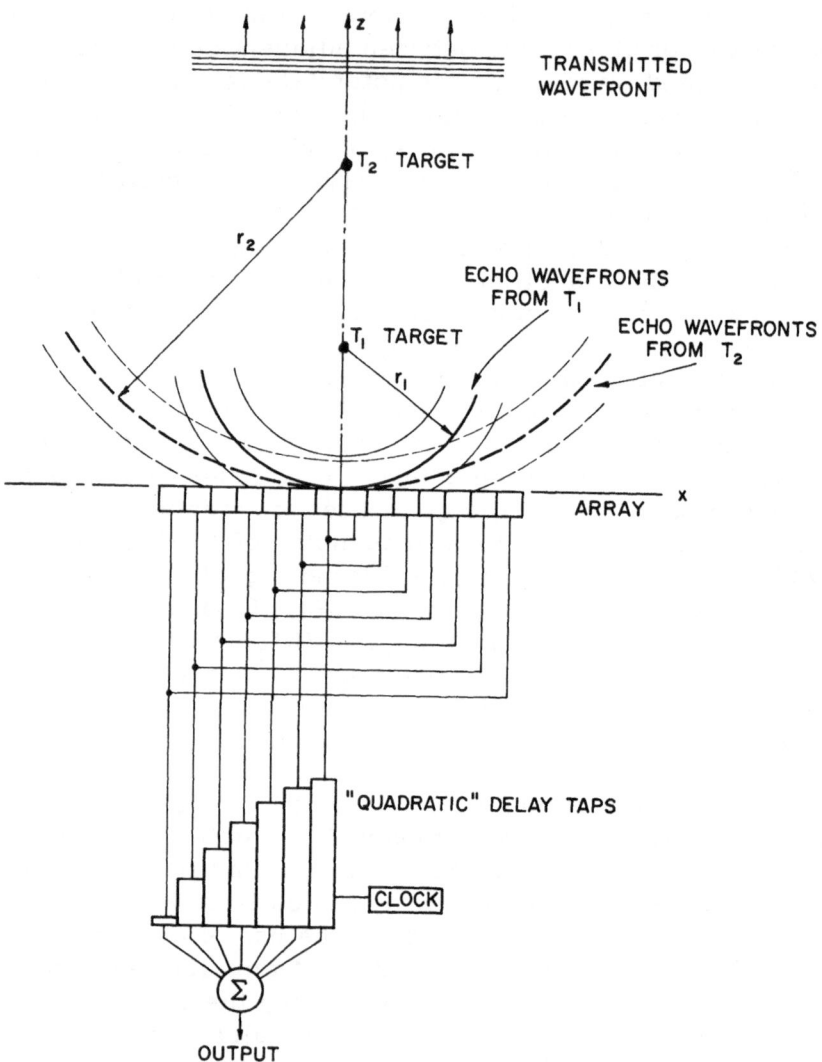

Fig. V-7. Dynamic focusing with quadratically tapped
           variable delay line (after Melen)

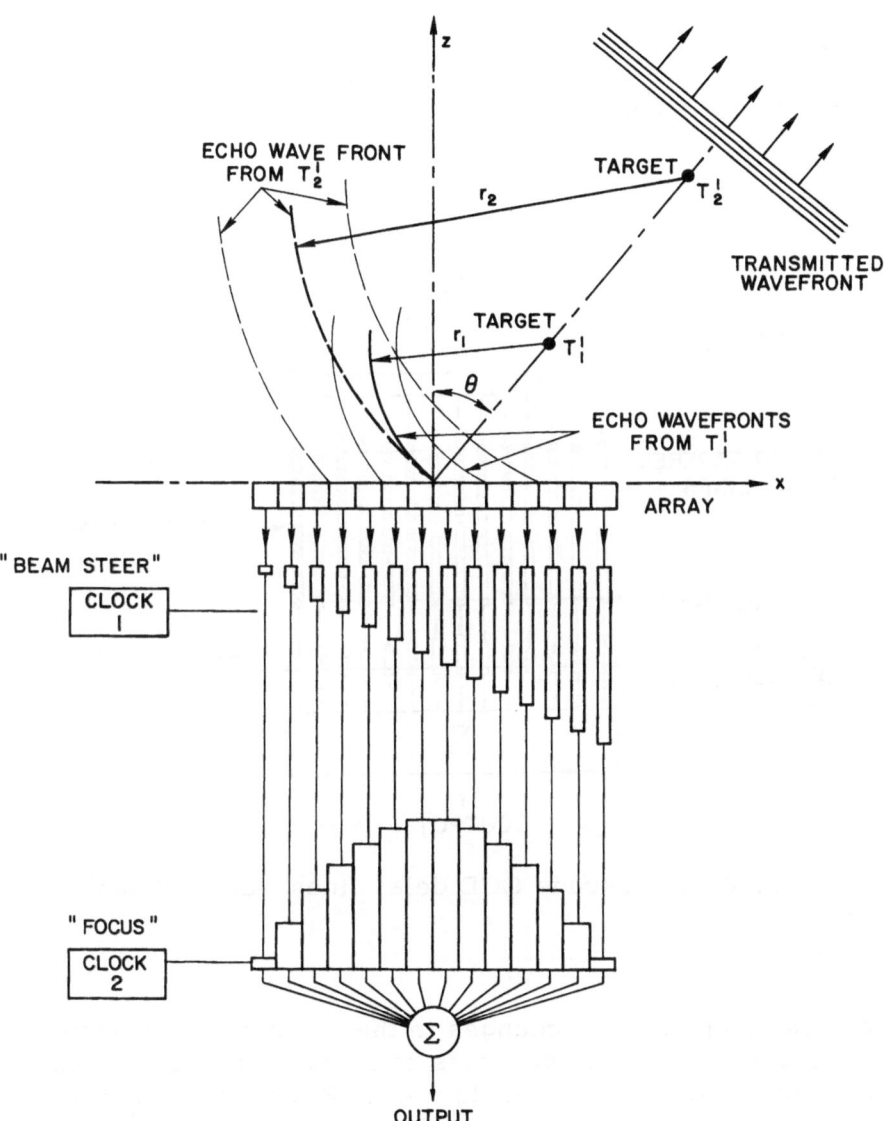

Fig. V-8. Dynamic focusing and beam steering with variable
delay lines (after Melen)

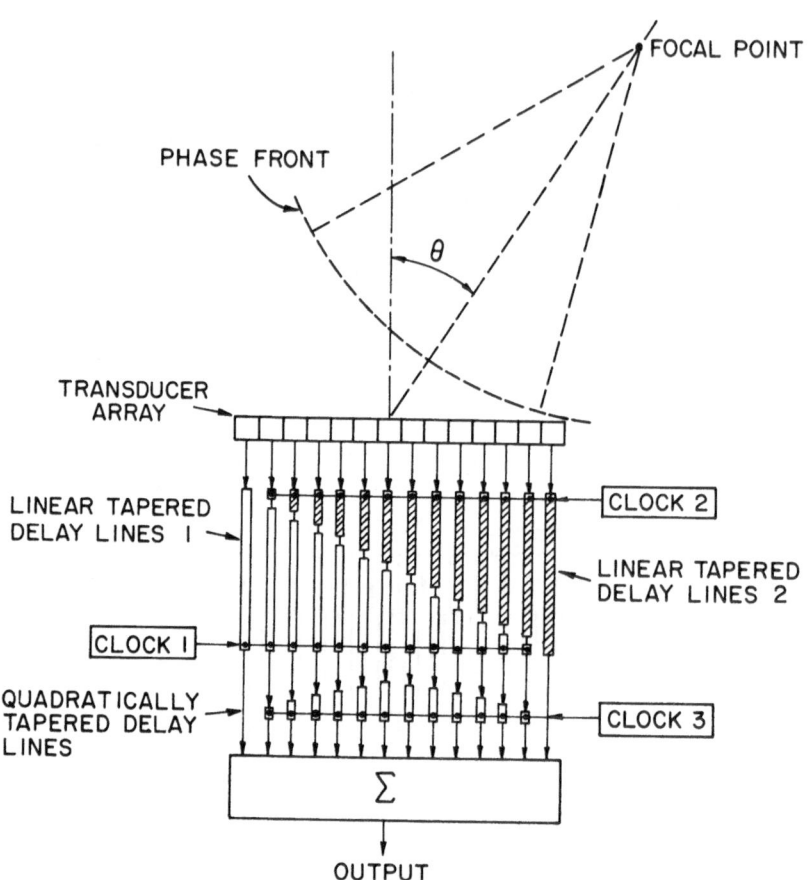

Fig. V-9. Cascade CCD delay lines (after Melen)

z-axis by appropriate changes in the two clock frequencies. For example, if the frequencies of clock 1 (f_1) and clock 2 (f_2) are equal, the scan angle $\theta = 0$. However, if $f_2 \ll f_1$, the beam is steered toward the far right, and if $f_1 \ll f_2$, the steering angle is toward the extreme left. As suggested by Fig. V-9, the three CCD arrays can be fabricated within a single silicon integrated circuit to form a CCD "electronic lens."

The array illustrated in Fig. V-9 does not provide high lateral resolution in the direction orthogonal to the plane of the scan. The "θ-array" illustrated in Fig. V-10 may solve this problem [7]. Here the product of the field patterns of a linear receive-only array and an annular transmit-only array is used to obtain high lateral resolution in all directions. In this system arrays of CCD delay lines are used in conjunction with both the receive and transmit transducer arrays. Compared with acoustic focusing, electronic focusing offers a most substantial improvement in depth of focus.

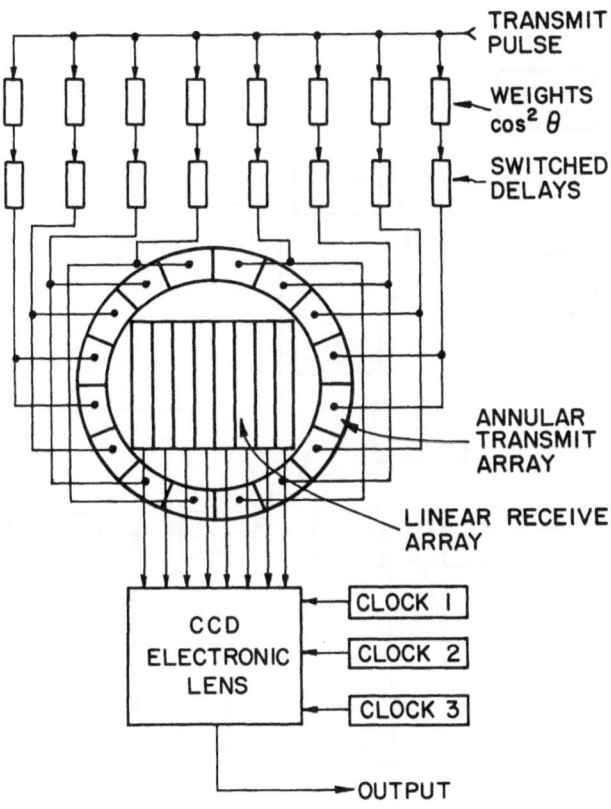

Fig. V-10. Electronically focused θ-array (after Macovski)

C. Holographic

The use of a 20 x 20 area array of receive-only transducers for holographic reconstruction of acoustic images is
illustrated in Fig. V-11 [10]. A mechanical stepper switch
is used in this experimental system to time share a single
signal processing channel among all elements of the transducer array. A parallel signal processing channel per transducer element and an extremely high speed digital computer
system would permit real-time operation of this system.

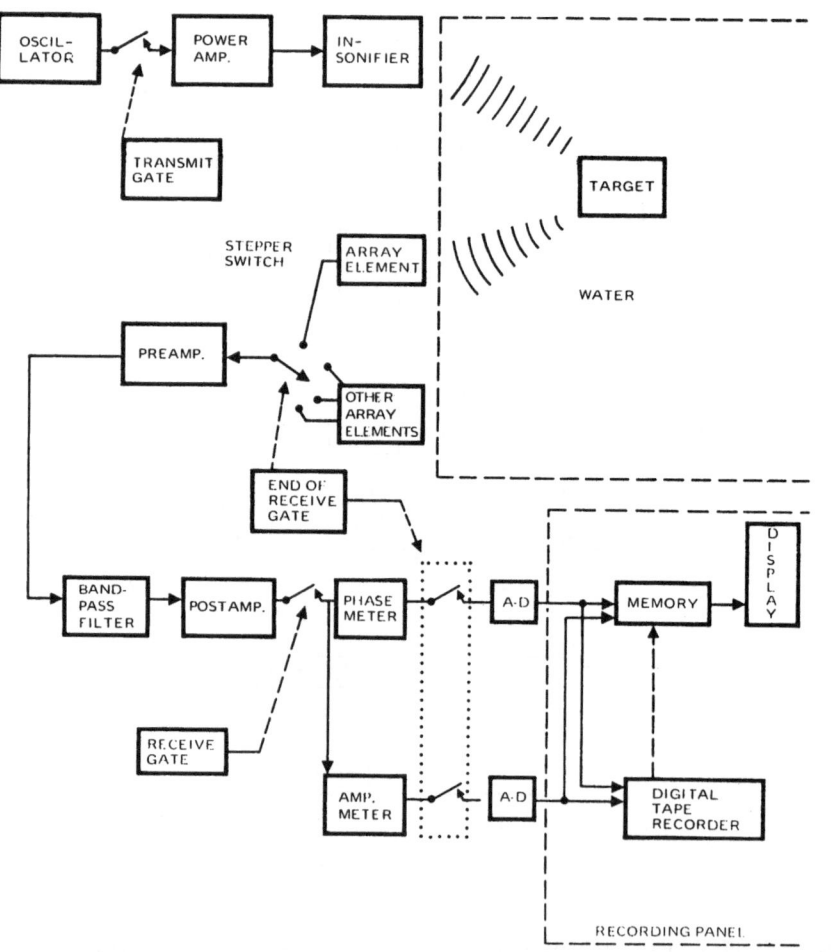

Fig. V-11. Control panel signal flow diagram (after Thorn)

VI.   CONCLUSIONS

The principal objective of this discussion has been a review of the salient electronic device and circuit advances which have served to implement the recent development of acoustic imaging arrays. The primary conclusions which are apparent following this review are:

1) arrays of well matched high voltage transmitter circuits are of considerable importance, particularly in transmit/receive electronic scanning or focusing systems;

2) arrays of well matched low noise, wide dynamic range preamplifiers and logarithmic compressors are of major importance in virtually all systems;

3) arrays of well matched quadrature mixers and high speed A/D converters are of special importance in holographic systems;

4) arrays of well matched high voltage, high current, low resistance multiplex switches are of major importance in transmit/receive systems;

5) arrays of well matched, large time delay-bandwidth product, wide dynamic range electronically variable analog delay lines are of dominant importance in electronic scanning or focusing systems. Charge coupled device (CCD) delay lines fabricated as silicon monolithic integrated circuits appear to offer a remarkably promising solution to the delay line problem; and

6) arrays of uniform batch-fabricated electrical inter-connections to the individual transducer elements are of major importance in two-dimensional area arrays.

In view of the needs for arrays of well matched trans-mitters, preamplifiers, logarithmic compressors, mixers, multiplex switches, CCD delay lines and electrical inter-connections enumerated above, the potential for monolithic integrated circuits in array systems would appear to be quite promising. Moreover, the peculiar requirements which medical imaging systems impose on these circuit arrays strongly suggest the utilization of custom integrated electronics.

Considering the performance capabilities of array systems in general, it appears that high volumetric resolution and a large volumetric field of view in real time are among the most difficult imaging system requirements to satisfy. The use of two-dimensional area arrays provides a promising approach to meeting these requirements.

The significant advances in transcutaneous ultrasonic imaging systems which have been achieved during the past several years through the use of transducer arrays offer great promise of even more substantial future progress.

VII.   ACKNOWLEDGEMENTS

The author gratefully acknowledges the valuable contributions of Drs. William Beaver, Albert Macovski, Maxwell Maginness, James Plummer and Roger Melen of the Stanford Electronics Laboratories, and Drs. Richard Popp, Leslie Zatz and Jason Birnholz of the Stanford Medical Center, to this discussion.

VIII.   REFERENCES

Section I

[1] Acoustical Holography, Vol. 4, G. Wade, Ed., Plenum Press, New York, 1972.
[2] Acoustical Holography, Vol. 5, P.S. Green, Ed., Plenum Press, New York, 1973.
[3] Wells, P.N.T., Physical Principles of Ultrasonic Diagnosis, Academic Press, New York, 1969.
[4] Meindl, J.D., "Integrated Electronics in Medicine," ISSCC Digest of Tech. Papers, Feb. 1974, pp. 160-161.

[5] Vilkomerson, D., "Analysis of Various Holographic Imaging Methods for Medical Diagnosis," Acoustical Holography, Vol. 4, op.cit., pp. 201-429.

Section II

[1] Wells, P.N.T., Physical Principles of Ultrasonic Diagnosis, Academic Press, 1969.
[2] Bom, N., et al., "Multiscan Echocardiography," Circulation 48:1066-1074, November 1973.
[3] Thurstone, F.L., and von Ramm, O.T., "A New Ultrasound Imaging Technique Employing Two-dimensional Electronic Beam Steering," Acoustical Holography, Vol. 5, op.cit., pp. 249-259.
[4] Maginness, M.G., Plummer, J.D., and Meindl, J. D., "An Acoustic Image Sensor Using a Transmit-Receive Array," ibid., pp. 619-631.
[5] Kossoff, G., "Display Techniques in Ultrasound Pulse Echo Investigations: A Review," J. Clinical Ultrasound 2(1):61-71, March 1974.

Section III

[1] Wells, P.N.T., Physical Principles of Ultrasonic Diagnosis, Academic Press, 1969.
[2] Ekoline 20 Manual, Smith Kline Corporation, Palo Alto, California.
[3] Wells, P.N.T., "The Receiver in the Pulse-Echo System," Ultrasonics in Medicine, M. de Vlieger et al., Eds., Excerpta Medica/American Elsevier Publishing Co., Inc. New York, 1974, pp. 30-37.
[4] Mussman, H., Dutton, R., and Meindl, J., "A Monolithic Analog Signal Processor for Ultrasonic Imaging Systems," ISSCC Digest of Tech. Papers, February 1975, pp. 180-181.
[5] Mussman, H., Integrated Circuits Laboratory, Stanford University, private communication.
[6] Griffith, J.M., and Henry, W.L., "A Sector Scanner for Real-time Two-dimensional Echocardiography," Circulation 49:1147-1152, June 1974.

[7] Eggleton, R.C., and Johnston, K.W., "Real-time Scanning System Compared with Array Techniques," Proc. IEEE Ultrasonics Symposium, Nov. 1974, pp. 16-18.

[8] Holasek, E., Sohollu, A., and Parnell, E.W., "A Digitized, Direct Contact B-scanner for Ophthalmic Application," J. Clinical Ultrasound 1:36-40, March 1973.

[9] Krause, W.E.E., and Soldner, R.E., "Ultrasonic Imaging Technique (B-scan) with High Image Rate for Medical Diagnosis," Digest 7th International Conference on Medical and Biological Engineering, Stockholm, Sweden, August 1967, p. 21-3.

Section IV

[1] Thurstone, F.L., and Melton, H.E., "Biomedical Ultrasonics," IEEE Trans. on Indus. Elec. & Control Inst. IEC-1(2):167-172, 1970.

[2] Burckhardt, C.B., et al., "Methods for Increasing the Lateral Resolution of B-scan," Acoustical Holography, Vol. 5, op.cit., pp. 391-413.

[3] Burckhardt, C.B., et al., "Focussing Ultrasound Over a Large Depth with an Annular Transducer-An Alternative Method," IEEE Trans. on Sonics & Ultrasonics SU-22: 11-15, January 1975.

[4] Green, P.S., et al., "A New High-Performance Ultrasonic Camera," Acoustical Holography, Vol. 5, op.cit., pp. 493-503.

[5] Fenner, W.R., and Stewart, G.E., "An Ultrasonic Holographic Imaging System for Medical Applications," ibid., pp. 481-492.

[6] Bom, N., et al., "Multiscan Echocardiography," Circulation 48:1066-1074, November 1973.

[7] Somer, J.C., "Electronic Sector Scanning for Ultrasonic Diagnosis," Ultrasonics 6(3):153-159, 1968.

[8] Somer, J.C., et al., "Ultrasonic Tomographic Imaging of the Brain with Electronic Sector Scanning System," Proc. IEEE Ultrasonics Symposium, Nov. 1974, pp. 43-48.

[9] Thurstone, F.L., and von Ramm, O.T., "A New Ultrasound Imaging Technique Employing Two-dimensional Electronic Beam Steering," op.cit.

[10] Lobdell, O.O., "A Nonlinearly Processed Array for Enhanced Azimuthal Resolution," IEEE Trans. on Sonics & Ultrasonics SU-15:202, Nov. 1968.

[11] Havlice, J.F., et al., "A New Acoustic Imaging Device," Proc. IEEE Ultrasonics Symposium, Nov. 1973, pp. 13-18.

[12] Fraser, J., et al., "A Two-dimensional Electronically Focused Imaging System," Proc. IEEE Ultrasonics Symposium, Nov. 1974, pp. 19-23.

[13] Vilkomerson, D., "Acoustic Imaging with Thin Annular Apertures," Acoustical Holography, Vol. 5, op.cit., pp. 283-316.

[14] Knollman, G.C., et al., "Linear Receiving Array for Acoustic Imaging and Holography," ibid., pp. 647-658.

Section V

[1] Cook, R.L., "Experimental Investigation of Acoustic Imaging Sensors," IEEE Trans. on Sonics & Ultrasonics SU-19(4), October 1972.

[2] Takagi, N., et al., "Solid-State Acoustic Image Sensor," Acoustical Holography, Vol. 4, op.cit., pp. 215-236.

[3] Harrold, S.O., "A Solid-State Ultrasonic Image Converter," Department of Electrical Engineering, Portsmouth Polytechnic, Portsmouth, England, Nov. 1972.

[4] Maginness, M.G., Plummer, J.D., and Meindl, J.D., "An Acoustic Image Sensor Using a Transmit-Receive Array," op.cit.

[5] Plummer, J.D., Meindl, J.D., and Maginness, M.G., "An Ultrasonic Imaging System for Realtime Cardiac Imaging," ISSCC Digest of Tech. Papers, Feb. 1974, pp. 162-163.

[6] Beaver, W.L., Maginness, M.G., Plummer, J.D., and Meindl, J.D., "Ultrasonic Imaging Using Two-dimensional Transducer Arrays," Cardiovascular Imaging and Image Processing: Theory and Practice- 1975, SPIE, Palos Verdes Estates, California (in press).

[7] Macovski, A., and Norton, S., "High-resolution B-scan Systems Using a Circular Array," Acoustical Holography, Vol. 6, N. Booth, Ed., Plenum Press, New York, 1975, pp. 121-143.

[8]   Melen, R.D., Shott, J.D., Walker, J.T., and
      Meindl, J.D., "CCD Dynamically Focused Lenses for
      Ultrasonic Imaging Systems," 1975 CCD Applications
      Conference, San Diego, California.
[9]   Barbe, D.F., "Imaging Devices Using the Charge
      Coupled Concept," Proc. IEEE 63(1):38-67, Jan. 1975.
[10] Thorn, J.V., Booth, N.O., Sutton, J.L., and Saltzer, B.A.,
      "Test and Evaluation of an Experimental Holographic
      Acoustic Imaging System," Naval Undersea Center,
      San Diego, California, NUC TP 398, November 1974.

Chapter 9

BRAGG-DIFFRACTION IMAGING

G. Wade

Electrical Engineering Department

University of California at Santa Barbara

9.1 INTRODUCTION

The first good images by means of Bragg-diffraction of laser light from sound were produced in 1966 by Korpel [1]*. Korpel's work followed the experimental conclusions reported in 1965 by Cohen and Gordon on the nature of light diffraction by ultrasonic wave fronts [4].

As noted in Chapter 2, the early pioneer in acoustic imaging, S. J. Sokolov, in one of his early attempts to detect flaws with ultrasound, used a bulk system which was remarkably similar in overall appearance to the modern Bragg-diffraction systems. One of the differences was that Sokolov's system employed Debye-Sears diffraction rather than Bragg diffraction. Also, in Sokolov's system no actual images were formed. The light pattern on Sokolov's screen (See Fig. 2 of Chapter 2) showed only the intensity and the number of orders of the diffracted sound. If the metal piece being tested were highly non-homogeneous, the sound passing through the acoustic cell would be greatly scattered and damped, and the intensity and number of orders on the screen

───────────────

* Although the system concepts were first enunciated and experimented with by Korpel, the ideas involved were independently conceived by Hance, Parks and Tsai [2] at Lockheed Research Laboratories and the present author and his students [3] at UCSB. As pointed out in a footnote in Chapter 2, research efforts similar to Korpel's at Zenith were mounted at about the same time by these workers.

would be small. For a homogeneous piece, the intensity and
number of orders would be high. Without a substantial
refinement of the system no actual images on the screen
could possibly have been formed.

As mentioned in Chapter 2, the techniques employed in
a Bragg-diffraction system are, strictly speaking, non-
holographic in nature. Nevertheless, phase information is
conserved in the Bragg-diffracted light that leaves the
sound cell. This fact suggests a way of looking at Bragg-
diffraction imaging that is reminiscent of holography.
Consider the simplest possible thick, optical hologram,
one in which the surfaces of maximum exposure are a series of
equidistant, parallel planes extending throughout the bulk
photographic material. Such a hologram would result if, in
recording it, we were to use both a planar object beam and
a planar reference beam. In the reconstruction process, we
would illuminate the hologram with a replica of the original
reference beam. Such a beam would automatically enter the
hologram at the Bragg angle with respect to the recorded
fringe pattern. The reconstructed object beam would then be
a   replica of the original planar object beam. A very
similar process can take place if we illuminate properly an
acoustic cell filled with water and containing the simplest
possible ultrasonic wave-front configuration, one that
involves a planar wave of sound progressing through the cell
as depicted in Fig. 1. If we irradiate the cell with a
planar wave of laser light that enters at the Bragg angle
with respect to the sound wave fronts, the Bragg-diffracted
light will also be planar and we will reconstruct a light-
beam replica of the original planar sound beam. The image
available from this diffracted light could then be thought
of as being an image of the planar transducer.

Thus we see that, in the case of simple planar waves,
Bragg diffraction from an acoustic cell can be interpreted
in the same way as Bragg diffraction from a thick hologram.
For more complicated situations, this is also true. We may
conceptually regard any ultrasonic beam whatsoever as consisting of a simple summation of plane-wave components
having different directions of propagation. This is true,
for example, of a sound beam that is scattered from an
object. Assume we use a laser to illuminate a sound cell
containing an object beam. Assume further the entering
laser light has a wide, uniform spatial spectrum, that is,
the light has a continual distribution of planar-wave

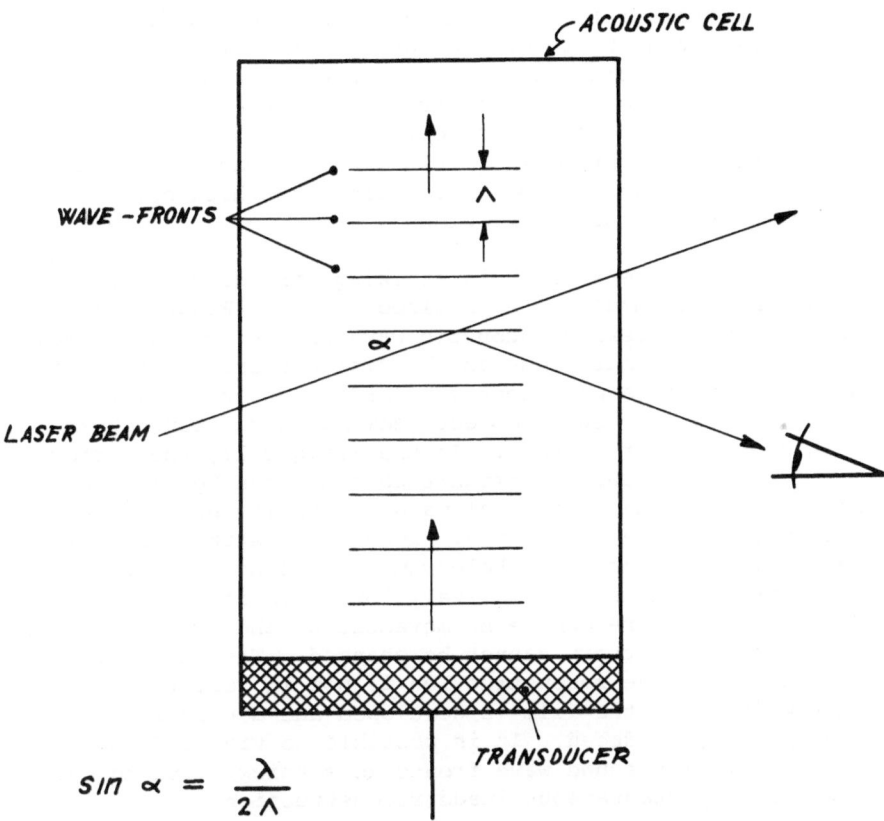

Fig. 1 Planar-wave reconstruction by Bragg diffraction of
laser light from an acoustic cell. The Bragg condition is given by the equation at the bottom of the
figure where λ is the wavelength of the light and Λ,
the wavelength of the sound.

components, all of equal amplitude and spread over a wide
range of propagation directions. Then the emerging diffracted light will possess component propagation vectors corresponding to each of those in the sound beam. The amplitude
of each diffracted component in the light will be proportional to that of the associated one in the sound. Thus
there will be a one-to-one correspondence between the
acoustic propagation vectors and the diffracted optical
propagation vectors. The diffracted light will in this
fashion constitute a sort of optical replica of the acoustic
object beam. By processing this light we can obtain an
image of the object. The analog with the thick optical
hologram is obviously very close.

Although the analogy is striking, the differences are
important and should be emphasized. In the Bragg cell, it
is a moving pattern of acoustic wave fronts from an object
beam by itself that produces the diffraction. In the thick
hologram, it is a permanently recorded pattern of interference fringes between an object beam and a reference beam
that does the diffracting. In the Bragg cell, the pattern
formation is virtually instantaneous and can be altered rapidly at will, the individual wave fronts moving through the
cell with the speed of sound. Image reconstruction is in
real time. In the thick hologram, recording the pattern
frequently requires a relatively long time of exposure
during which there can be no movement of the fringes. Once
recorded, the pattern cannot be changed. The record is
permanent. Image reconstruction is not in real time but
must wait until the film is developed and a holographic
transparency produced. It is possible to view a Bragg cell
with its moving sound wave fronts as a thick, live hologram,
capable of instantaneous image reconstruction.

A Bragg-diffraction imaging system is, above all,
simple. It does not require complex special processing
procedures to form the image. It can produce real-time
images using either a pulsed mode of operation or continuous
waves. In principle, it can instantaneously construct
optical replicas of any sound field whatsoever retaining
phase information as well as amplitude information. In
practice, there are important limitations on how precise the
replicas can be. It happens, for example, that in ordinary
Bragg-diffraction imaging it is very difficult to reproduce
the acoustic propagation vector distributions, as described
above, in all three dimensions. To do this would impose

detrimental limitations on the object's position, the image
resolution, etc. [5]. Therefore, in almost all systems of
this type, the propagation-constant replication process
previously described operates in one plane only. The images
of object components running in one direction are produced
as stated, but those of object components running at right
angles to this direction are generated in a conceptually
more complicated fashion [6], [7]. For this reason, the
character of the resolution and the aberrations in the image
are different in the two directions [8].

9.2 DESCRIPTION OF THE SYSTEM

A diagram of a conventional Bragg-imaging system is
shown in Figure 2. The beam from a helium-neon laser is
introduced through a diverging lens and is spread to an
appropriate diameter. At this point, a collimating lens is
placed so as to make the laser beam parallel. The beam is
then passed through a cylindrical lens to form a converging
wedge of light which enters the acoustic cell through a glass
window.

The cell is a water-filled box with a quartz transducer mounted at one end. The transducer is driven by an
RF generator-and-amplifier combination set to the desired
frequency. The transducer produces in the water a beam of
longitudinal acoustic waves which travel at 1500 m/sec
through the region illuminated by the converging light beam
and thence to the opposite end of the cell. For imaging by
transmission, the object to be imaged is placed in the
acoustic beam in the area between the transducer and the
converging light. This results in a complex distribution
of scattered acoustic waves. Since the waves are essentially
pressure waves, they produce correspondingly small fluctuations in the refractive index of the water as predicted by
the Lorentz-Lorenz law. These local variations cause the
laser light to be diffracted in a manner first conjectured
by Brillouin [9]. Since the effective width of these scattered waves is typically equal to many times their wavelength,
there is in this acoustic cell, an angular selectivity in
the direction of the diffracted light components that is
analogous to that found in the well-known Bragg diffraction
which takes place when certain crystals are radiated with
X rays. This phenomenon makes Bragg-diffraction imaging
possible.

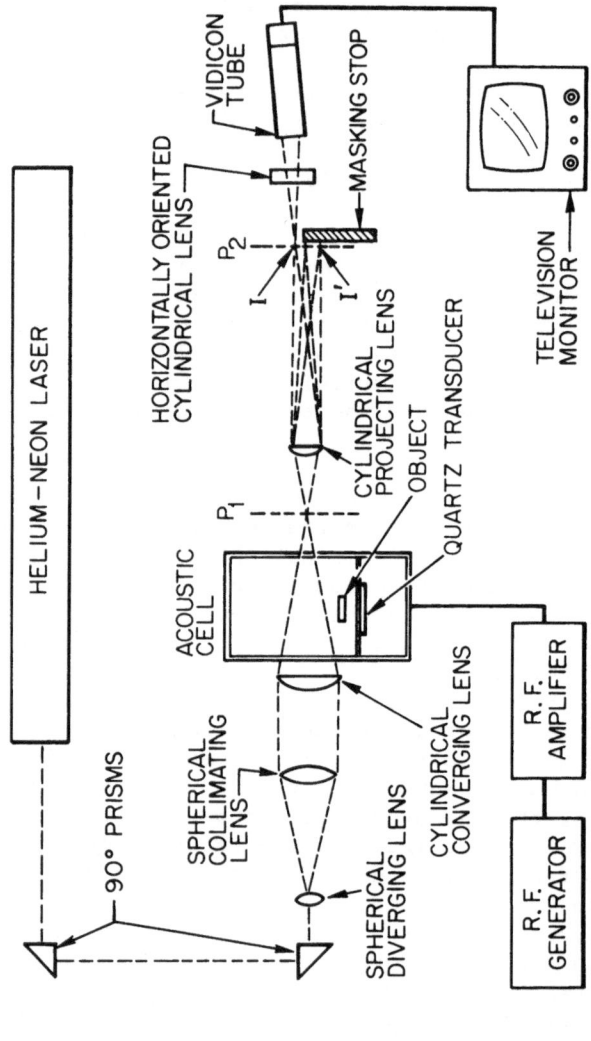

Fig. 2 Diagram of a Bragg-diffraction imaging system.

The diffracted light produces two images located on either side of, and very close to, the line focus for the undiffracted light. These images are available at plane P_1 (shown in Fig. 2) for processing by optical components. The images are carried by an optical representation of the sound beam but they are distorted to the extent that, while the magnification in one dimension is unity, the magnification in the orthogonal dimension is given by the factor λ/Λ where λ is the optical wavelength and Λ the acoustic wavelength. For typical values of acoustic frequency, this factor is equal to about 1/100 or less.

The distortion referred to above is due to the fact that, as previously explained, the process for replicating the propagation constants operates only in one plane, that is, the plane of the paper on which Fig. 1 is drawn.

This distortion is overcome by projecting the images onto plane P_2 with a cylindrical lens of short focal length. A stop is used in plane P_2 to allow only one of the beams to proceed. This beam is then passed through still another cylindrical lens whose axis is rotated 90° with respect to those of the first two cylindrical lenses. This last lens is used to correct any distortions in aspect ratio which still remain in the image. In many of the systems, the image beam then falls onto the face of a vidicon television tube and eventually appears on the screen of a television monitor. In less sophisticated systems the image beam may simply be projected onto an observation screen at the image plane. The fundamental process in the image formation is that of Bragg diffraction of the laser light from the scattered sound. It is the treatment of this acousto-optical interaction which will concern us in the following two sections of this chapter.

9.3 THE LIGHT-SOUND INTERACTION

The Bragg-diffraction imaging technique can be most readily understood by first considering the interaction between a monochromatic, planar beam of acoustic waves and a monochromatic, planar beam of light. If the acoustic beam is sufficiently wide (for truly planar waves the width is infinite), it will be found that the light, in general, passes through the sound without any sustained diffraction taking place. This is a result of the effects of destructive interference among the various diffraction products in the light which tend to be generated by interaction with

the sound. However, for certain specific angles of incidence of the light, the acoustic wave fronts behave in a
manner analogous to the atomic planes in a crystal when
illuminated by X rays at the so-called Bragg angles. For
these particular angles, a significant amount of light is
diffracted in a sustained fashion by the acoustic wave fronts,
as was previously discussed in connection with Fig. 1. A
constructive combination of all the diffraction products
generated in the light leads to the buildup of the Bragg-
diffracted light beam.

The Bragg angles can be determined from geometrical
considerations, but it is equally valid and, to the person
familiar with quantum mechanics, simpler to consider it
from the point of view of conservation of momentum in a
photon-phonon interaction. However, because of limited
space I shall not derive the Bragg condition here. The
expression for the Bragg angles is given by Eq. (1).

$$\sin\alpha = \frac{n\lambda}{2\Lambda} \tag{1}$$

where λ is the optical wavelength, Λ is the acoustic wavelength and n is any integer.

For weak interactions, it is reasonable to consider
only the case where n = 1. Making this assumption for the
value of n, we find that there is just one angle of incidence
at which plane, monochromatic light waves will be diffracted
by plane, monochromatic acoustic waves.

Of course, in the case of Bragg imaging, the scattered
acoustic field is no longer a simple plane-wave column, but
rather some complicated distribution of waves traveling in
many directions. It is possible, however, to conceptually
treat this or any other distribution as an infinite summation
of plane-wave components similar to the treatment of time
signals in communications electronics. This approach
requires a spatial analog to the temporal Fourier transform
with which electrical engineers, schooled in communications
theory, are very familiar.

Thus the analytical approach being used here is one
consistent with a communications-theory point of view. This
particular approach for studying optical and sonic systems
is proving to be an invaluable tool. It is not by accident
that electrical engineers with a background in communication
systems have spearheaded the present remarkable advances in

holography and in the general area of optical data processing. Analytical approaches, previously applied to problems
involving the time-frequency domain of electronics, are now
being used in the spatial domain of optics and sonics with
great effectiveness. The next section will describe the
technique of imaging by ultrasonic Bragg diffraction of
laser light in terms of modern communications theory.

9.4   SPATIAL FOURIER TRANSFORMS

When using Fourier transforms to treat electronic
signals, the transforms are always functions of a single
variable, either time or frequency. For application of
the transform theory to spatial field distributions (optical
or acoustic, as in the case of Bragg imaging) we must generally be concerned with functions of the two variables
corresponding to the orthogonal directions of the coordinate
system used to describe the cross section of a particular
field. Thus, if we have a scalar field $U(x,y)$, the field
can be represented by the Fourier integral which is shown
in Eq. (2).

$$U(x,y) = \int\int_{-\infty}^{\infty} U'(f_x,f_y) e^{+j2\pi(f_x x+f_y y)} df_x df_y \qquad (2)$$

where f_x and f_y are "spatial frequencies" and $U'(f_x,f_y)$
represents the amplitude of each component plane wave.
While these spatial frequencies may seem confusing at first,
they are really nothing more than a specification of direction of propagation of the particular plane-wave component.
In fact, they are simply equal to the direction cosines
divided by the wavelength of the propagation vector representing the traveling waves.

We know from Fourier-transform theory that the component amplitudes can be represented by the following transform:

$$U'(f_x,f_y) = \int\int_{-\infty}^{\infty} U(x,y) e^{-j2\pi(f_x x+f_y y)} dx dy \qquad (3)$$

The advantage of using such transforms in analyzing
imaging systems is primarily that it enables us to take
advantage of the mathematical structure involving transforms which have been developed for communications theory.

This becomes more clear if we consider the case of an
object irradiated by plane acoustic waves. In Fig. 3, the

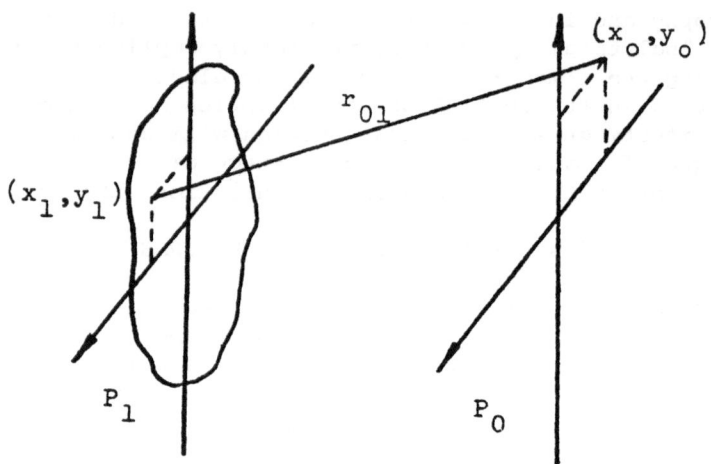

Fig. 3   Acoustic diffraction by an aperture.

representative object is an aperture in a plate which is
otherwise opaque to sound. The plate is located at the
plane P_1. Assume that the plate is illuminated from behind
by acoustic waves of wavelength Λ and we wish to find an
expression for the wave distribution at some observation
plane P_0. Using the Rayleigh-Sommerfeld diffraction formula
we can write [10]

$$U(x_0,y_0) = \iint_{\text{aperture}} U(x_1,y_1)\frac{1}{j\Lambda}\frac{e^{jkr_{01}}}{r_{01}}\cos(\overline{n},\overline{r}_{01})dx_1dy_1$$

(4)

where $r_{01} = \sqrt{z^2+(x_0-x_1)^2+(y_0-y_1)^2}$. The so-called "obliquity
factor", $\cos(\overline{n},\overline{r}_{01})$, can be considered approximately equal
to unity for this situation. The expression can be simpli-
fied by using the Fresnel approximations for diffraction,
and we obtain

$$U(x_0,y_0) = \frac{e^{jkz}}{j\Lambda z}e^{j\frac{k}{2z}(x_0^2+y_0^2)}\int\limits_{-\infty}^{\infty}\int U(x_1,y_1)$$

$$e^{j\frac{k}{2z}(x_1^2+y_1^2)}e^{-j\frac{2\pi}{\Lambda z}(x_0x_1+y_0y_1)}dx_1dy_1.$$

(5)

Upon close examination, it becomes evident that the integral
can be identified as the spatial Fourier transform of the
function

$$U(x_1,y_1)e^{j\frac{\pi}{\Lambda z}(x_1^2+y_1^2)}$$

where the spatial frequency f_x is given by $x_0/\Lambda z$ and f_y, by
$y_0/\Lambda z$.

Thus we find that the amplitude distribution at the
plane P_0, assuming that it is in the Fresnel-diffraction
region, is given by

$$U(x_0,y_0) = (\frac{e^{jkz}}{j\Lambda z}e^{j\frac{k}{2z}(x_0^2+y_0^2)})(F.T.\{U(x_1,y_1)e^{j\frac{k}{2z}(x_1^2+y_1^2)}\})$$

$$(6)$$

where F.T. stands for "Fourier transform of." If, for
convenience, we confine ourselves to the consideration of
an object whose dimensions are quite small with respect to
Λz, then the factor

$$e^{j\frac{k}{2z}(x_1^2+y_1^2)} = e^{j\frac{\pi}{\Lambda z}(x_1^2+y_1^2)}$$

produces only a slight "phase curvature." For purposes of
physical understanding, little is lost if we consider this
factor to be equal to unity. Then it is evident that the
acoustic-wave amplitude distribution produced by a scattering object is proportional to the spatial Fourier transform of the object itself (recalling that the incident
radiation is of constant amplitude over the aperture and
thus is constant for all values of x_1 and y_1, which fall
within the region of the integration).

This point is clarified if we consider a simple object
such as a narrow slit in an opaque screen. Such a slit is
the spatial analog of the square pulse which we often
encounter in electronics, and the Fourier transform of a
square pulse is proportional to the well-known sinc function. So we would expect the transform for the slit to be
similar. If we consider the slit as being of infinite
length (but, of course, of finite width), then the spatial
transform reduces to one dimension and is precisely of the
sinc form. For the electronic square-pulse signal, the

transform variable in the sinc function represents the
temporal frequency. Thus the function provides the frequency
spectrum corresponding to the component sine waves in the
pulse. For the case of a slit producing acoustic-wave
diffraction, the transform variable represents the spatial
frequency of the waves in the plane of observation. Since
the acoustic energy is monochromatic, the wavelength is a
constant; therefore, the spatial Fourier transform is in
terms of the direction of propagation of the component waves.
For this reason, the transform is often referred to as an
"angular spectrum."

The Bragg-diffraction imaging technique depends upon
the ability of the system to form an optical replica of the
acoustic-wave pattern. As we have previously discussed,
an object is irradiated with sound. The sound, scattering
from the object, then passes into the interaction region.
By means of Bragg diffraction, this scattered sound causes
various light components to be diffracted out and away from
an otherwise converging incident laser beam which enters into
the same interaction region. As mentioned before, we assume
that the entering light has a wide, uniform spatial spectrum.
Then the emerging diffracted light will possess component
propagation vectors corresponding to each of those in the
scattered sound beam. If we operate in the linear regime,
the magnitude of each component in the Bragg-diffracted light
will be proportional to the magnitude of the component in
the sound which does the diffracting. Thus the image
information is transferred from the scattered sound to the
Bragg-diffracted light.

If the spectrum of the entering light is not uniform
in amplitude, then the Bragg-diffracted light will not be
a precise replica of the scattered sound. The magnitude
of any particular component in the Bragg-diffracted light
is not only proportional to that of the corresponding
component in the sound, but also to the magnitude of the
incident light component which is being diffracted. For
example, if no component exists in the incident light
which meets the Bragg condition with respect to a particular
one in the scattered sound (that is, if the magnitude of
that light component is zero), there will be no corresponding
component in the Bragg-diffracted light (that is, its magnitude will also be zero). Thus we can conclude that as
far as the entire output spectrum is concerned, the spatial

spectrum (and hence the spatial Fourier transform) of the
Bragg-diffracted image-forming beam of laser light is
proportional to the spatial spectrum (and hence the spatial
Fourier transform) of the incident light multiplied by the
spatial spectrum (and hence the spatial Fourier transform)
of the scattered sound. This principle was first enunciated
with mathematical rigor by Korpel [11], and obviously corresponds to a convolution in the space-variable domain.

9.5 FILTERING CAPABILITY OF A BRAGG-DIFFRACTION SYSTEM

From the above, it is obvious also that the spatial
spectrum of the incident light can be thought of as a
transfer function $H(f_x, f_y)$. In communication-systems theory,
a transfer function is defined as the ratio of the transforms of the output and input signals. A similar definition
can be used for acousto-optical systems. Fig. 4 illustrates
the situation diagrammatically and emphasizes the fact that

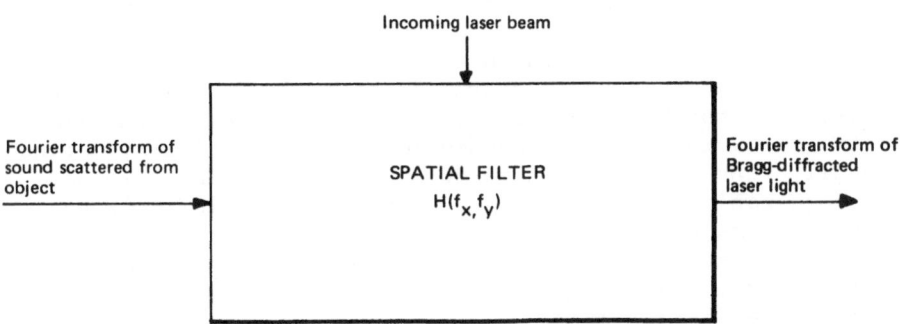

Fig. 4   Diagram illustrating how the incoming laser beam
          serves as a spatial filter in the optical-beam
          replication process.

the incoming laser beam serves as a spatial filter for the
system. From this we can conclude that the transform (or
angular spectrum) of the incident light must be uniform
over all spatial frequencies for the Bragg-diffraction
image-forming beam of light to be an exact replica of the
scattered acoustic beam.

We can readily understand the nature of such a beam of
incident light if we consider the equivalent case in electronics. A signal whose transform is uniform is simply one
that contains all frequencies in equal proportion -- an
impulse. For the one-dimensional imaging case which we are
considering, the optical analog of an impulse is a line
source, or more specifically, the cylindrically converging
wave fronts which produce a line of illumination.

It is not convenient in practice to illuminate the
acoustic beam with truly cylindrical light waves converging
to a line over a full spectrum of 360°. We compromise by
using instead the partial spectrum associated with the
converging wedge of light previously described and shown
in Fig. 2. Such a wedge contains only those angular spectral components which fall within the angle of convergence.
Because of this, the Bragg-diffraction system behaves as a
band-pass filter.

The above analogy is very useful because it enables
us to determine the nature of the image which will result
from a particular imaging system. In electronics, when a
square pulse is fed through a band-pass filter, the shape
of the pulse is modified in a way which we can predict
by knowing the filter's characteristics. The same is true
here. As the bandwidth is narrowed, the edges in the image
of a slit, for example, become less and less clearly defined
until, with a very small bandwidth, the slit is no longer
recognizable. The need for a wide bandwidth is important
for the Bragg system just as it is for a radar receiver,
where the accuracy with which we can determine the position
of a target depends upon the ability of the receiver to
resolve a narrow pulse.

The concept of filtering can be used also in the design
of special Bragg-diffraction imaging systems in which some
particular imaging characteristic is desired [12]. For
example, assume there is need for a system which will
accentuate the high spatial frequencies in an object. High

spatial frequencies correspond, of course, to fine detail.
For such a purpose it would be possible to use an incident
light beam which behaves as a high-pass filter. Such a beam
might take the form of a converging, hollow wedge of light
oriented in the proper direction. The beam would not furnish components for Bragg diffraction corresponding to those
of low spatial frequency in the scattered sound. Thus low-
frequency components in the image light would be eliminated;
only high-frequency components would be present.

From the above discussion, we can see that the input
light field of the Bragg-diffraction imaging system has a
role analogous to that of a filter transparency in an optical spatial-filtering system. The effect of the plane-wave
components in the scattered sound that ultimately produce
the image beam of light can be selectively changed by modification of the incident light field.

The spatial-filtering capability of a Bragg-diffraction
imaging system can be illustrated by considering a couple of
images obtained several years ago by an early embodiment of
this type of system. Fig. 5 shows a silhouette or transmission image of a hook made of copper wire having a diameter of 1.0 mm. The physical orientation of the wedge
of light with respect to the acoustic cell for this image
was such that the low-frequency spatial components in the
scattered sound beam (that is, those components corresponding
to essentially unscattered sound coming directly from the
insonifying transducer) were strongly Bragg-diffracted and
appear with maximum intensity in the replicated, diffracted
laser light. For making this image, the Bragg angle (α
of Fig. 1) was quite small. Therefore to achieve the above
condition the axial plane of the cylindrical converging lens
had to be oriented almost perpendicularly with respect to
the long dimension of the acoustic cell, very nearly as
depicted in Fig. 2.

Fig. 6 is an image of the same hook produced with an
arrangement for eliminating the low-frequency spatial
components and accentuating the high spatial frequencies.
To do this, the cell of Fig. 2 was rotated with respect to
the wedge of light in that figure. Imagine that the cell
is rotated about an axis extending out of the paper. The
cell eventually reaches an orientation (as illustrated in
Fig. 7) such that no light can be diffracted by any unscattered component of the sound. The result of this, of course,

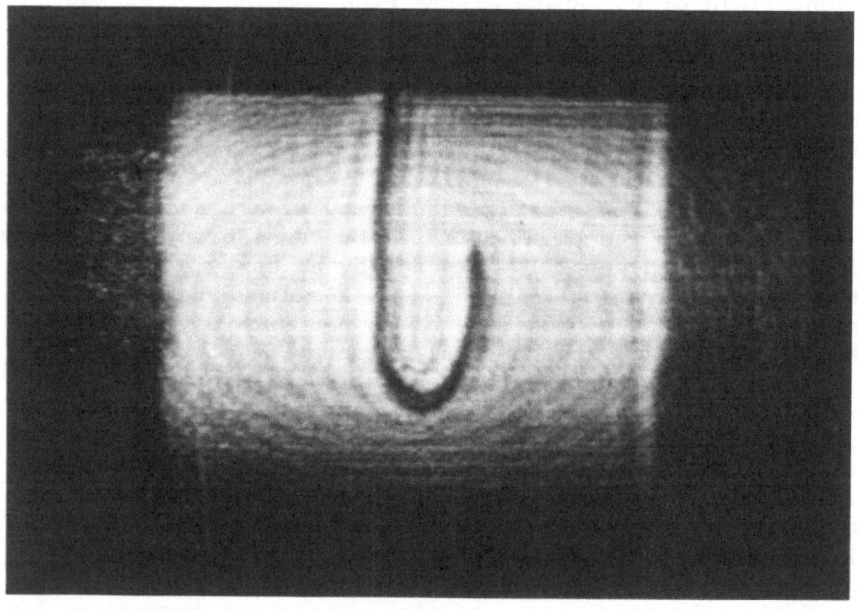

Fig. 5   Transmission image of a wire hook when the low-
         frequency spatial components have been retained.

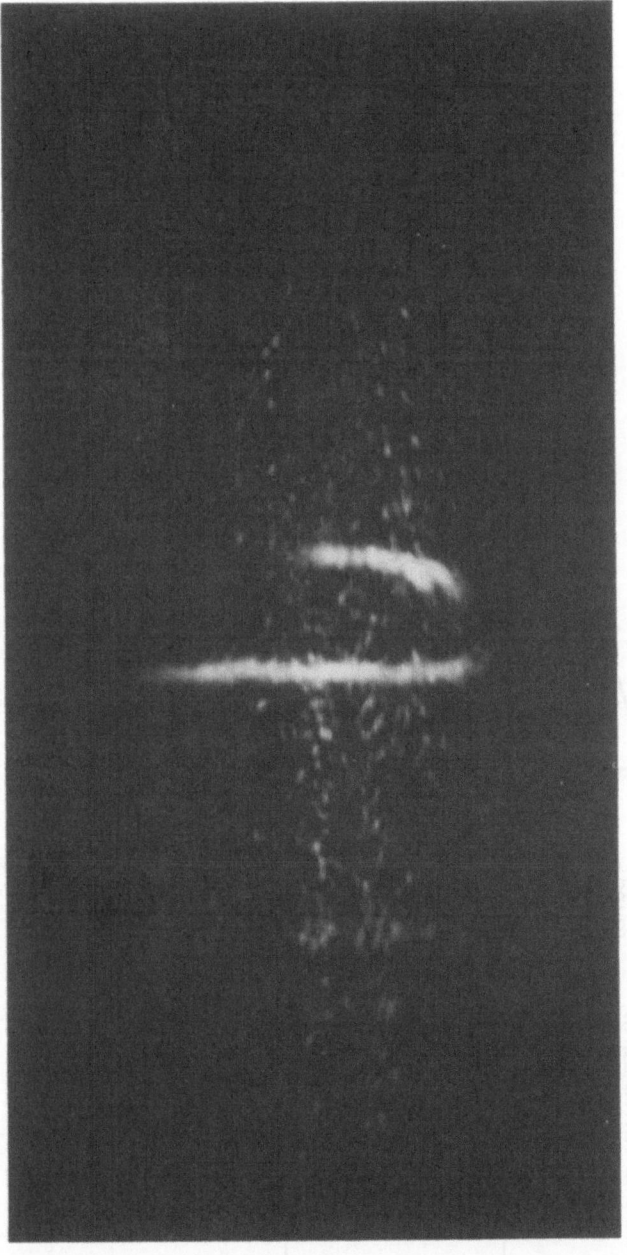

Fig. 6 Transmission image of hook when the low frequency spatial components have been eliminated and the high-frequency components, accentuated.

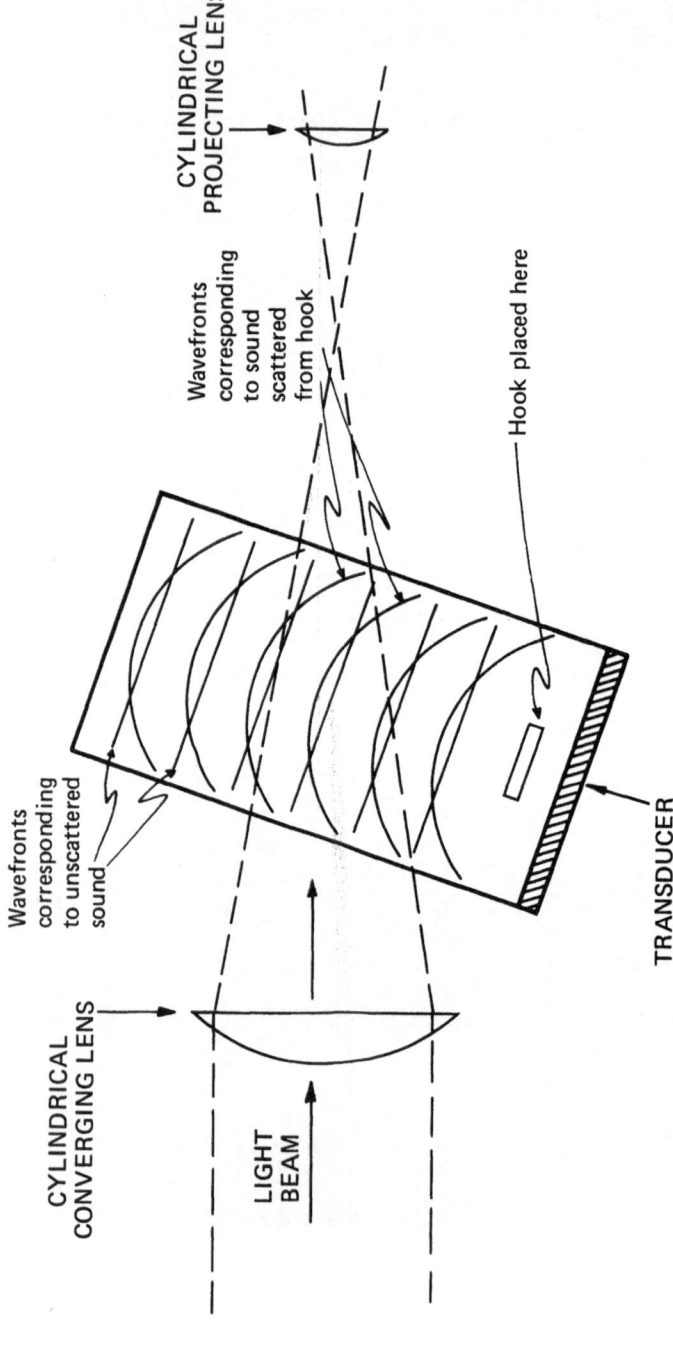

Fig. 7 Diagram illustrating cell rotation with respect to the light-beam wedge in obtaining the image of Fig. 6. Note that this method of acoustic imaging is analogous to the dark-field microscopy of optics.

is that no bright background, corresponding to the acoustic-beam cross section, can appear on the image screen. If now the object is placed in the usual position for transmission imaging (that is, between the transducer and the light wedge), there will be no diffracted light in the image beam corresponding to the central-order sound component. Thus, in the image of Fig. 6, the bright background (corresponding to low-spatial frequencies in the sound) has been eliminated. However, the higher spatial frequencies, corresponding to the sound components scattered through large angles from the hook, are included, and even accentuated, in the image. This causes the brightness of the hook itself to be enhanced as illustrated in the figure.

9.6   STRENGTHS AND WEAKNESSES OF BRAGG-DIFFRACTION IMAGING

Bragg diffraction is perhaps the simplest of the real-time orthoscopic imaging methods, and by and large, uses less costly components. Besides a laser and a sound cell, the only other components used in the conventional system are two spherical lenses for collimating the laser beam, two or three cylindrical lenses for focusing it, and an observation screen for viewing the image. In practice, however, the viewing is usually accomplished by using a television camera and monitor (see Fig. 2). As will be pointed out later, the cylindrical optical lenses for focusing the image can be eliminated by inserting a cylindrical acoustical lens into the sound cell between the object and the interaction region.

In principle, a Bragg-diffraction system can instantaneously construct optical replicas of any sound field whatsoever, retaining phase information as well an amplitude information. Nevertheless, in practice, a number of difficulties are frequently encountered. Although Bragg systems operating in the vicinity of 20 MHz produce excellent images, special problems arise in going to the lower frequencies. A frequency as high as 20 MHz is perhaps suitable for non-destructive testing, but for medical diagnosis the lower frequencies with their higher penetrating capability are necessary. At a frequency as low as 3 MHz, the Bragg-diffraction angle of Fig. 1 is less that 1 milliradian. With such a small angle, the undiffracted laser readout beam (the zero-order beam) and the two first-order image beams will be physically located very close together. The optical image at this frequency will receive a much larger part of the extraneous light from the zero-order beam than in the case of the higher-frequency operation where the Bragg angle is much larger. This extraneous light with its high quantum noise (as described in Chapter 4) constitutes the largest source of unremoveable noise in the system, and thus severely limits the inherent sensitivity of a Bragg-diffraction imaging system.

Even at the higher frequencies, the conventional Bragg-diffraction systems are not very sensitive. As explained in Chapter 4, all systems using laser-beam readout tend to have lower sensitivity than similar systems in which the readout is accomplished piezoelectrically.

9.6.1 Improving Sensitivity by Holographic Detection of Images

The frequency of the image-forming light in a Bragg-diffraction system is shifted from that of the light origi-nally fed into the system by an amount equal to the acoustic frequency. On the other hand, most of the background light, which acts as the major source of noise in the image, remains at the original frequency. This fact suggests that substan-tial noise reduction, and hence sensitivity improvement, can be realized if the image is derived from the frequency-shifted light by means of a process which eliminates the effects of the unshifted light. In a preliminary experiment, D.C. Winter has shown that this is so [13].* The image was produced by making a hologram whose object beam consisted of the frequency-shifted, Bragg-diffracted image light and whose reference beam had a frequency precisely equal to that of the image light.

Soo-Chang Pei and the author of this Chapter have analyzed this imaging technique in terms of acoustic thresh-old contrast and sensitivity [14]. The analysis shows that all light at the original frequency is greatly suppressed, as far as the holographic recording is concerned. The effect of this light as a noise source is down by an aston-ishing 11 orders of magnitude for a typical set of param-eters. Thus this source of noise is no longer predominant in limiting the sensitivity. Under idealized conditions, the sensitivity improvement due to this fact amounts to about three orders of magnitude. The analysis is presented

---

* The same idea was independently conceived and recorded by H. Keyani, et al., of UCSB in 1970.

in the following paragraphs.

As stated above, flare from the zero-order undiffracted
light constitutes the major source of noise in limiting the
sensitivity of the conventional Bragg-diffraction imaging
system [15]. Because of the finite aperture of the cylindrical converging lens of Fig. 2, the incident light is
scattered to both sides of the line focus into the image
location, thus producing the flare which degrades the image.
This invasion of unwanted light can not be removed by
spatial filtering. However, the holographic approach mentioned
above can be applied to the system in order to solve this
problem [13]. An optical hologram is made of the acousto-
optic image by utilizing a reference beam having a frequency
equal to that of the image light. The setup is shown in
Fig. 8.

Since the image and reference beams are of the same
frequency, their interference pattern consists of stable
fringes which can be recorded on film. (Note that because
film exposure and development are involved, the system under
these circumstances no longer operates in real time.) Due
to the frequency difference between the noisy flare light
and the reference light, the fringes produced by their
interference are not stationary. The individual fringes of
this latter pattern will move a distance of one fringe
spacing in a period of time equal to $1/\Omega$, where Ω is the
difference between the flare light and reference frequencies
(Ω is also the acoustic frequency). If the film exposure
time is significantly greater than $1/\Omega$, the moving fringes
will be averaged out and for all practical purposes will
not be recorded. When the holographic image is reconstructed
in the usual manner, only a duplicate of the acoustic image
is seen, free of the background noise from the flare light.
The purpose of the following paragraphs is to analyze this
system in terms of acoustic threshold contrast and sensitivity. By treating the hologram as a temporal filter [16],
we can easily show why this technique is so efficient in the
noise-reduction process.

9.6.1.1 The Hologram as a Temporal Filter. Before
considering the details of the system, we should first
explain the temporal filtering property of a hologram.
Fig. 9 shows the configuration of the holographic recording.
A photographic plate is illuminated simultaneously by a

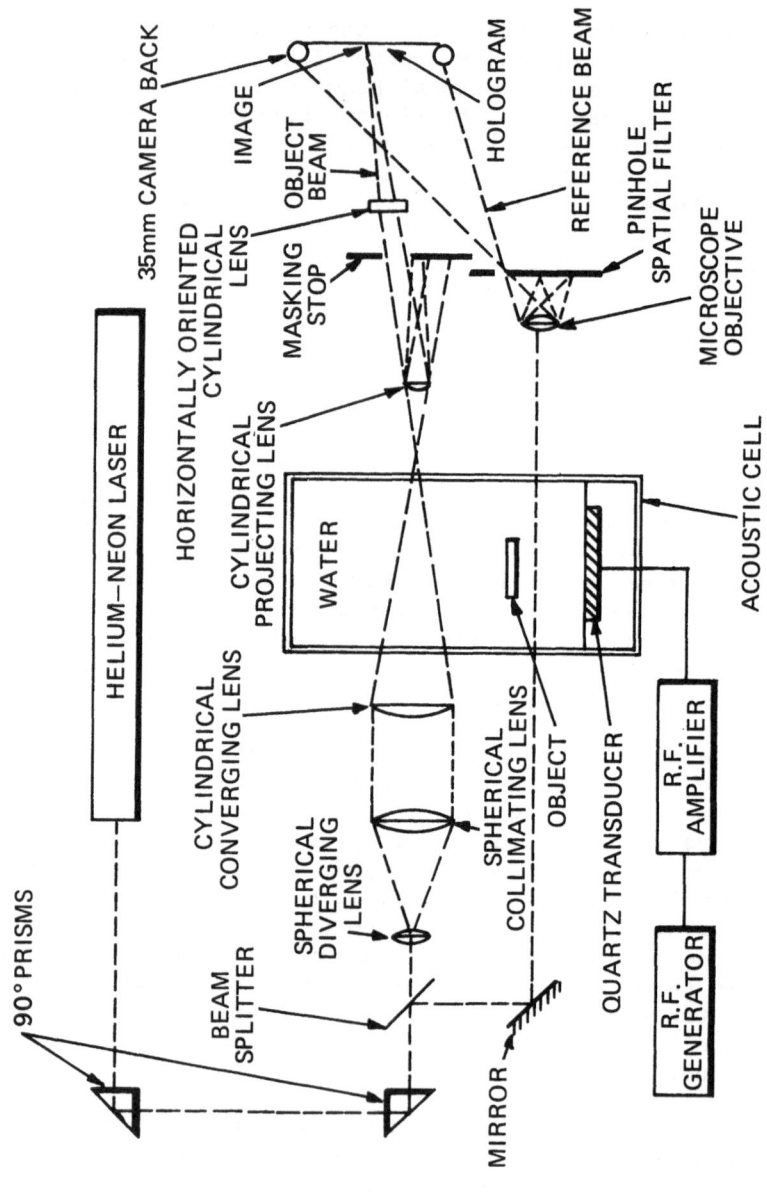

Fig. 8 Diagram of the Bragg-diffraction system with holographic recording.

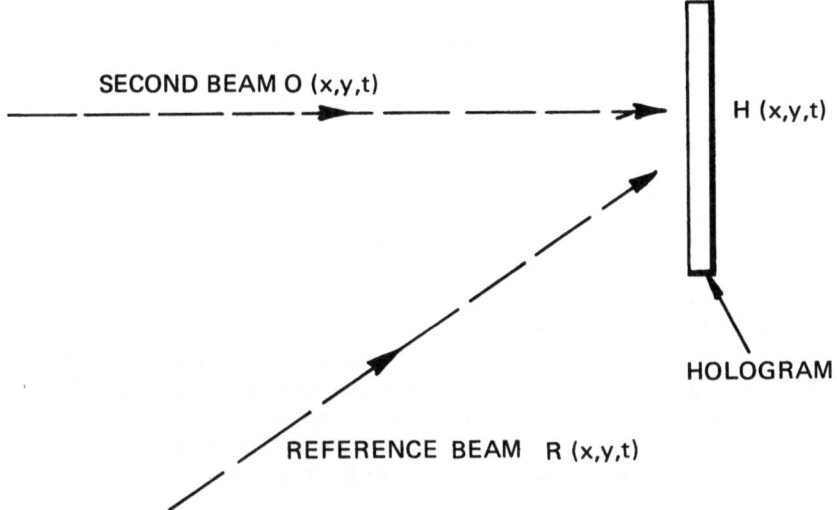

Fig. 9 General configuration of holographic recording.

monochromatic planar reference beam of frequency ν_0 and a
second beam at the slightly different frequency ν. We can
write the exposure at any point (x,y) on the film as

$$E(x,y) = \int_{-\frac{T}{2}}^{\frac{T}{2}} |H(x,y,t)|^2 dt \qquad (7)$$

where $|H|^2$ is the intensity distribution across the film
and T is the exposure time. We can write H as

$$H(x,y,t) = R(x,y,t)+O(x,y,t) \qquad (8)$$

Hence the exposure is

$$\dot{E}(x,y) = E_R + E_0 + \int_{-\frac{T}{2}}^{\frac{T}{2}} R(x,y,t) 0^*(x,y,t) dt$$

(9)

$$+ \int_{-\frac{T}{2}}^{\frac{T}{2}} R^*(x,y,t) \ 0(x,y,t) dt$$

where E_R and E_0 are the individual exposures due to the reference beam and the second beam respectively. In a conventional hologram the other two terms would be responsible for generating the twin images characteristic of the wave-front reconstruction process. We will therefore restrict our attention to the last term which would lead to the virtual image. Let us call this term $E_{V.I.}$. Thus

$$E_{V.I.} = \int_{-\frac{T}{2}}^{\frac{T}{2}} R^*(x,y,t) \ 0(x,y,t) dt$$

(10)

For simplicity we have assumed that the reference beam is a monochromatic planar wave having the optical frequency ν_0. (Actually, the planar-wave assumption does not affect the derived conclusions.) Hence

$$R(x,y,t) = U_r \ \exp(+j \ 2\pi\nu_0 t)$$

(11)

By combining Eqs. (10) and (11) we obtain

$$E_{V.I.} = U_r^* \int_{-\infty}^{\infty} [\text{rect}(t/T) \ \exp(-j2\pi\nu_0 t)] 0(x,y,t) dt$$

(12)

where

$$\text{rect}(t/T) = \begin{cases} 1 & , \ -\frac{T}{2} \le t \le \frac{T}{2} \\ 0 & , \ \text{otherwise} \end{cases}$$

(13)

let $\tilde{0}(x,y,\nu)$ be the Fourier transform of $0(x,y,t)$ defined by

$$\tilde{0}(x,y,\nu) = \int_{-\infty}^{\infty} 0(x,y,t) \ e^{-j2\pi\nu t} dt$$

(14)

We may use Parseval's theorem [17] to write Eq. (12) as

$$E_{V.I.} = U_r^* T \int_{-\infty}^{\infty} Sinc[T(\nu-\nu_0)] \overset{\sim}{0}(x,y,\nu) d\nu \qquad (15)$$

where $T\ Sinc[T(\nu-\nu_0)] = \dfrac{Sin[\pi T(\nu-\nu_0)]}{\pi(\nu-\nu_0)}$ and is the Fourier

transform of $[rect(t/T)exp(j2\pi\nu_0 t)]$.

From Eq. (9) we can see that the effect of a uniform time exposure of duration T is entirely equivalent to that of a linear filtering operation with a temporal transfer function $H(\nu) = Sinc[T(\nu-\nu_0)]$. This is illustrated in Fig. 10.

The time-exposed hologram therefore acts as a temporal band-pass filter. If the optical frequency of the radiation incident on the film from a given source (such as the zero-order flare light) is ν' cycles/second different from the optical frequency of the reference beam, the amplitude of the corresponding image point will be suppressed by a factor $Sinc(T\nu')$ as depicted in Fig. 10.

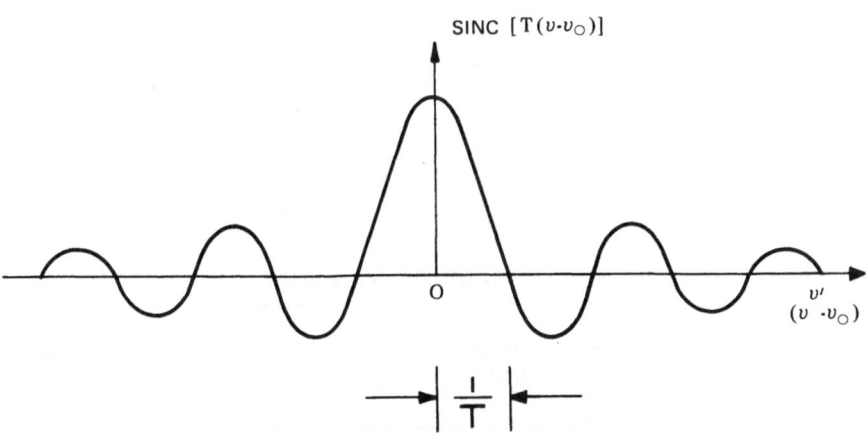

SINC $[T(\nu-\nu_0)]$

Fig. 10   The temporal transfer function of a time-exposed hologram.

Consider a noisy beam that illuminates the recording plane, the frequency of this beam being different from that of the monochromatic reference wave. The temporal filtering property of the hologram can obviously be used in the fashion indicated above to reduce the effect of such a beam. The noise which would otherwise degrade the image is thus suppressed by the coherent wave-front reconstruction process. Only those signals lying in a narrow frequency band of approximate width 2/T cycle/second about the reference frequency ν_0 (as shown in Fig. 10) will contribute substantially to the image. Therefore this holographic technique will strongly improve the performance of the system by virtually eliminating the effect of the zero-order light as illustrated in the block diagram of Fig. 11.

9.6.1.2 <u>Noise Limitations in Bragg-Diffraction Imaging</u>. As described above, the noise characteristics of a conventional Bragg-diffraction imaging system can be improved by temporal holographic filtering. Consider now the noise situation in detail. The image-forming light which acts as a signal in a Bragg-diffraction imaging system is frequency-shifted from the incoming light by an amount equal to the acoustic frequency. On the other hand, most of the background light which acts as noise against the desired image,

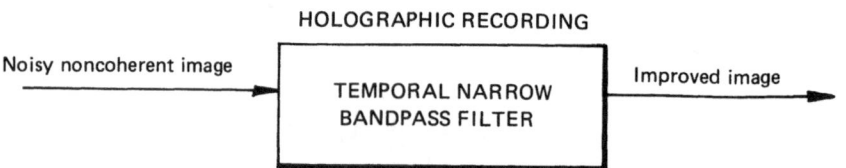

Fig. 11   The holographic technique acts as a temporal filter to improve the system performance.

remains at the frequency of the incoming light. The various
sources of image noise in the system include Tyndall scattering of light from particles suspended in the liquid,
Rayleigh scattering from refractive index inhomogeneities in
the liquid, zero-order flare light diffracted from the
finite lens aperture, and Brillouin scattering from thermal
phonons in the liquid [18]. Among these kinds of noise,
only the Brillouin scattering produces a shift in the frequency of the incoming light [18]. In the case of all the
other kinds, there is no shift. For low insonification, it
is the quantum noise due to the zero-order flare light (QNZ)
which predominates and limits the system performance [15].

As stated, because of the finite aperture of the cylindrical converging lens of Fig. 2, the incident light is
diffracted to both sides of the line focus into the image
location degrading the image. This is shown in greater
detail in Fig. 12. This invasion of the image location by
noisy flare light can not be removed by spatial filtering.
There are no optical filters which would completely separate
the two closely-spaced components (zero and first-order
Bragg-diffracted light) because of the low acoustic frequencies generally employed in a Bragg-imaging system,
especially one designed for biological imaging where the
acoustic frequencies must be small. Eq. (1) shows that
for a low frequency (high Λ) the Bragg angle will be small.
The amount of zero-order light scattered into the image
region can be computed from diffraction considerations and
imaging rules. Calculations [19] show that for diffraction-
limited resolution and small light-wedge angle, each vertical
column (column parallel to the line focus of the light wedge)
of resolution elements in the image is flooded by approximately one of the side lobes of the zero-order sinc distribution (see Fig. 12). For an example of optimum operation with
an object size of 10 cm x 10 cm, each of the resolution elements in the image is plagued by the power P_z from the zero-
order light [19], where

$$P_z = 1.25 \times 10^{-8}(1-s)P_0 \qquad\qquad (16)$$

In Eq. (16), P_0 is the laser power in watts, and s is the
Bragg-diffraction efficiency.

Results from previous work show that unless this zero-
order flare light can be eliminated from the image, the ideal
sensitivity of the system is limited by quantum noise due to

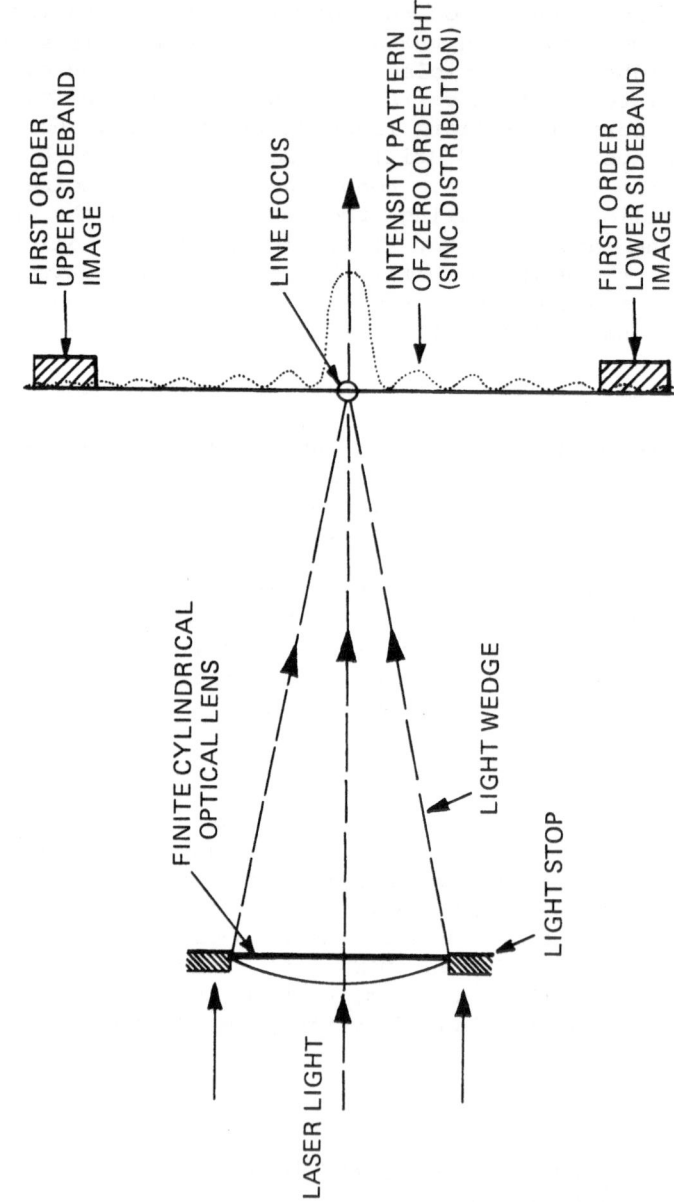

Fig. 12 Top view of the cylindrical converging lens and the light wedge it forms in a Bragg-diffraction imaging system.

the zero-order flare light scattered into the image region (QNZ).

9.6.1.3 Noise Reduction and System Improvement by Holographic Recording. From the preceding discussion, it is clear that we can take advantage of the fact that there is a frequency difference between the image and the above-described noise light. In Fig. 8, the reference beam is derived by passing a portion of the light coming from the laser through the Bragg cell between the transducer and the object. In this fashion, we can frequency-translate the reference beam so that its frequency differs from that of the original laser beam by an amount equal to the sound frequency. We employ a first-order temporal side band, that is, a Doppler-shifted beam of light having the same temporal frequency ($(\nu+\Omega)$ in Fig. 8) as the image light in the system. Consistent with the previous discussion, ν is the frequency of the laser light and Ω is the frequency of the sound. A hologram is made by the mutual interference between this monochromatic reference beam and the noisy object beam as shown in Fig. 8.

The temporal frequency ν of the unshifted noise light in the object beam is Ω cycles/second different from that of the upshifted reference beam ($\nu+\Omega$). Therefore the amplitude of the noise light will be suppressed by a factor sinc$(T\Omega)$, as previously described. It is interesting to calculate for a specific case how much the effect of the zero-order flare light can be reduced by the holographic recording. If we use 3 MHz for the acoustic frequency Ω, and 1/30 second for the typical exposure time T of the hologram, we obtain

$$\text{Bandwidth of filter of Fig. 10} \cong 2 \times \frac{1}{T} = 2\,\frac{1}{1/30} = 60 \text{ Hz}$$

The flare light operating point on the filter characteristic of Fig. 10 is given by

$$\frac{\Omega}{\frac{1}{T}} = \frac{3\times10^6 \text{ Hz}}{30 \text{ Hz}} = 10^5 \text{ sidelobes}$$

Thus the hologram is the equivalent of a temporal band-pass filter of bandwidth approximately 60 Hz. The temporal frequency of the zero-order flare light is located 10^5 side-lobes away from the central pass band (as defined by the

temporal sinc transfer function) of the hologram.   (See Fig.
10.)   The local maximum of the 10^5 sidelobe on a normalized
scale for the intensity is

$$\frac{1}{(10^5\pi)^2} \cong 10^{-11}$$

Hence the zero-order light is greatly suppressed (by 11
orders of magnitude) in the holographic recording. The original noise power level P_z is reduced by a suppression
factor of 10^{-11}.

$$P_z' = P_z \times 10^{-11} = 1.25 \times 10^{-8} \text{ (1-s) } P_0 \times 10^{-11}$$

(17)

$$P_z' = 1.25 \times 10^{-19} \text{ (1-s) } P_0$$

where P_z' is the equivalent noise power level for this
noise source using the holographic technique.

    To determine whether QNZ (the zero-order flare light)
still limits the performance, P_z' will be compared with the
noise due to Brillouin scattering into each image resolution cell. Call this latter noise P_B. From Reference [19]
we have

$$P_B = 1.83 \times 10^{-14} P_0$$

(18)

By comparing Eq. (17) with Eq. (18) we can see that P_B is
approximately 5 orders of magnitude larger that P_z',
since (1-s) is close to unity. Thus, quantum noise due to
Brillouin scattering (QNB) predominates over QNZ and limits
the system performance. From Reference [15] we have

$$C_{a\ QNB} = 0.6 \times 10^7 \text{ K}_0 \sqrt{\frac{\lambda\phi}{P_I}} \cdot \sqrt{\frac{h\Omega}{qPT}} \cdot \sqrt{\frac{5.32\Omega\alpha LKT_0}{I_s\Lambda^2}}$$

(19)

where $C_{a\ QNB}$ is the acoustic threshold contrast, K_0 is the
threshold signal-to-noise ratio for visual detection
(usually taken to have a numerical value of 5), λ is the
wave length in vacuum of the laser light in cm , ϕ is the
half-angle of the laser-beam wedge in radians, P_I is the
total laser power per unit height of the wedge in watts/cm,
h is Planck's constant, Ω is the acoustic frequency in

Hertz, Q is the quantum efficiency of the pick-up device, τ
is the integration time per resolution cell (1/30 second
film exposure for the photographic system), P is the power
in watts of the sound incident at one resolution element of
the object plane, L is the height of the light wedge in cm ,
K is Boltzmann's constant, T_0 is the room temperature in
water (taken to be 300°K), I_s is the acoustic intensity in
watts/cm^2 of the sound beam, and Λ is the acoustic wave
length in centimeters.

For a typical set of operating conditions, the minimum
detectable insonification intensity, $I_{s\ min}$, can be calculated from Eq. (19) by setting $C_{a\ QNB}$ equal to unity and
solving for I_s. Thus we obtain

$$I_{s\ min} = 7 \times 10^{-12} \tag{20}$$

where we have used λ = 4880Å, P_I = 0.2 watts/cm, θ = 1/10
radian, L = 10 cm, K_0 = 5, Q = 0.1, τ = 1/30 second, δA
(the area of a resolution cell) = 5/2 Λ^2, Ω = 3 MHz, and
$P = I_s \times \delta A$.

Now let us compare Eq. (20) with the previously
published I_{min} of a conventional Bragg-diffraction imaging
system. From Reference [15], we have I_{min} = 6.1 x 10^{-9}
watts/cm^2, where τ was taken to be 1/30 second. Thus, the
sensitivity has improved by about 3 orders of magnitude.

The above calculation obviously assumes that it is
now quantum noise due to Brillouin scattering that limits
the system performance. In a more complete analysis of
this holographic technique, we have also looked into other
sources of noise to see whether this assumption is correct.
For example, we have checked into the question of quantum
noise associated with the reference beam and also with the
object beam leaving out the flare light. We have shown that
for small amounts of acoustic power and for a laser which
has good coherence, QNB is the noise which puts the highest
lower limit on the minimum detectable insonification, and
hence gives the correct $I_{s\ min}$. Thus, under the conditions
assumed in this analysis, the predominant noise source
limiting the system performance is quantum noise in the
Brillouin scattering.

If we make similar calculations for other kinds of
systems we can see that the sensitivity of this system,

with its holographic detection of the image, compares quite
favorably with that of any other acoustic imaging system
[15]. However, it should be noted that in using the holographic approach we no longer have a system that operates
in real time. Nevertheless, with a frame time of only
1/30 second, it is obvious that we can record many holographic frames per second and, in effect, obtain a movie
which will display the original object motion. In this
mode of image viewing, not just one hologram is made of
the object, but a large number, each in rapid sequence.
These holograms can then be made to constitute the various
frames of a holographic movie. When the frames are put
together, the movie will reconstruct the object in its
precise original motion. An observer, using a simple holographic viewer, can either run the movie with the focus at
a selected plane, or he can stop the action to focus on
different depths within the object region in any one frame.
He has the option of being able to focus throughout the
entire volume using only a single hologram if he wishes to
do so.

9.6.2 Speckle And Ringing

In addition to problems of sensitivity, speckle and
ringing are also a problem. The system employs coherent
laser light and coherent sound. When the light is Bragg-
diffracted from the sound, the phase information associated
with the sound is retained in the diffracted light, as we
have seen. Because of this, as Chapter IV explains, the
images from a Bragg system usually contain much ringing
and speckle.

9.6.3 Cylindrical Lenses, Optical and Acoustic

Perhaps the most difficult problem encountered in
constructing a large-aperture Bragg-diffraction imaging system is that having to do with the quality of the optical components, particularly the cylindrical converging lens of
Fig. 2. Large-diameter spherical lenses, such as telescope
lenses or aerial-camera lenses, with good corrections for
aberrations, are readily available. However, cylindrical
lenses are quite another matter. The only large-scale
commercial application for high-quality cylindrical elements
is found in the anamorphic projection required by the motion-
picture industry. Unfortunately, the cylindrical lenses
employed in this application are free of aberrations only
when used in combination with spherical elements for which

they are designed. The cylindrical lenses in the Bragg-
diffraction systems, operating at the higher frequencies,
are not so critical a factor in the system operation because
the acoustic wavelength is small enough that a large wedge
angle (large numerical aperture) in the interacting light
beam is not so necessary as in the low-frequency case. At
the low frequencies where the wavelength is large, a high
numerical aperture is quite essential. In a low-frequency
system built at UCSB, the best cylindrical lens that could
be obtained with sufficient size to be used with the acoustic
cell, had a relative numerical aperture of only 0.05. The
resolution corresponding to this figure is 8.5 acoustic
wavelengths, or approximately 2.5 millimeters. Therefore
the images produced by that low-frequency system were not
of high quality.

Recent experimental and theoretical results have shown
that the above problem can be solved by inserting a cylindrical acoustic lens into the system between the object
and the light-beam wedge [20]. The resolution will then
depend upon the numerical aperture of this lens rather
than that of the optical lens. This is advantageous because
it is relatively easy and inexpensive to construct an acoustic
lens with a high numerical aperture. One reason for this
is the fact that materials with a wide range of indices of
refraction are available for making acoustic lenses. For
example, acoustic lenses can easily have a relative refractive
index ratio with respect to water of 4:1 or even higher;
optical lenses, on the other hand, have relative refractive
index ratios with respect to air of only about 1.5:1.
Another reason for the ease and low expense in constructing
an acoustic lens is that large dimensional tolerances are
allowable in an acoustic lens, as opposed to an optical lens,
because the wavelengths are large.

Let us now examine in some detail why employing a
cylindrical acoustic lens can increase the horizontal resolution of the system. In the conventional Bragg-diffraction
system, where no acoustic lens is used, the horizontal
resolution, as previously stated, is determined by the spatial
filtering action of the wedge of light which probes the
scattered sound beam. Assume that the object is a line
source of sound, the line extending in the same direction as
the apex of the wedge. The spatial spectrum of acoustic
rays emanating from the line source enters into the inter-

action region and encounters the laser light in the wedge. If the acoustic spectrum is wide and the wedge angle narrow, the light-wave spectrum will not be wide enough to provide the light rays needed for meeting the Bragg condition with respect to all the acoustic rays. This point has previously been explained. Under these circumstances only a portion of the sound rays will be able to scatter light (via Bragg-diffraction) into the image beam. Consequently the resolution will be degraded, the light wedge probing only a finite portion of the spectrum of the line source of sound. The diffraction-limited resolution therefore depends on the extent of the spectral window and hence on the numerical aperture (NA_O) of the wedge. Thus

$$\xi_h = \frac{\Lambda}{2\ NA_O} = \frac{\Lambda}{2\ \sin\ \phi} \tag{21}$$

where ξ_h is the resolution for object structure which varies in the horizontal direction (the direction parallel to the long dimension of the quartz transducer in Fig. 2), Λ is the acoustic wavelength and ϕ is the semi-apex angle of the light wedge.

The nature of the horizontal resolution can be changed with the insertion of a cylindrical acoustic lens in the sound cell as shown in Fig. 13. To illustrate the operation, assume that a line source is located at the front focal plane of the acoustic lens. This lens collects and collimates acoustic rays originating from the line source. The filtering of the acoustic spectrum in this case occurs at the acoustic lens, not at the light-sound interaction region. Hence the diffraction limited resolution is determined by the numerical aperture of the acoustic lens (NA_a). According to the Rayleigh resolution criterion

$$\xi_h = \frac{\Lambda}{2\ NA_a} = \frac{\Lambda}{2\ \sin\ \theta} \tag{22}$$

where 2θ is the angular aperture of the lens; that is, the angle subtended by the two extreme rays in the figure.

The insertion of the acoustic lens must be accompanied by the addition of a processor component in the form of an optical cylindrical lens. This latter lens is needed to perform an inverse transformation for viewing the image. We can see why such a transformation is necessary if we regard the acoustic lens as a Fourier-transforming device. (For

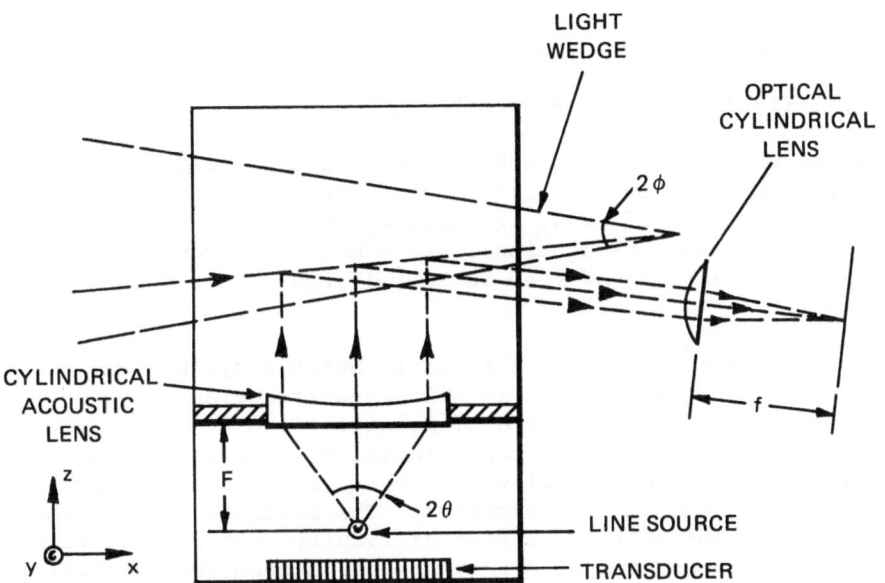

Fig. 13 Diagram of the acoustic cell of a Bragg-diffraction
         imaging system with an acoustic lens.

an object placed in the back focal plane of a convergent lens the pattern at the front focal plane is always the object's Fourier transform.) The inverse transform is then obviously needed to produce the image in its proper form.

Thus with the acoustic lens in place, the numerical aperture of the light wedge no longer affects the horizontal resolution. However, that numerical aperture does play an important role in defining the field of view. This is illustrated in Fig. 14. In the conventional system, a wide field of view is rather easily obtained by making the sound cell with its transducer wide, and the wedge long. By adding the acoustic lens to the system, the field of view may well be reduced sharply because it will then be given by

$$\text{Field of view} = F \cdot 2\phi = 2 \frac{r_\ell}{\sin \theta} \phi \tag{23}$$

where r_ℓ is half the width of the acoustic lens. Note that there is a trade-off relationship between horizontal resolution and field of view. To make ξ_h in Eq. (22) small (for good horizontal resolution), $\sin \theta$ must be large. However, to make the field of view in Eq. (23) large, $\sin \theta$ must be small.

Workers at UCSB have recently built a cylindrical acoustic lens to be used in a Bragg-diffraction imaging system. The lens was made of plexiglas and had a numerical aperture of 1/9.2. The experimental results with this lens in the system agree well with theory, the horizontal resolution being improved as expected. The acoustic lens in this fashion eliminates the severe difficulty with respect to an optical component in the conventional system as previously discussed.

There is yet another advantage to the use of the cylindrical acoustic lens. It can be seen from the geometry of Fig. 14 that if the distance between the image screen and the optical cylindrical lens is equal to the focal length of the acoustic lens, the aspect ratio of the image will be correct. In the conventional system illustrated by Fig. 2 this is not the case. An aspect ratio correction must be provided by the horizontally-oriented cylindrical lens shown in that figure.

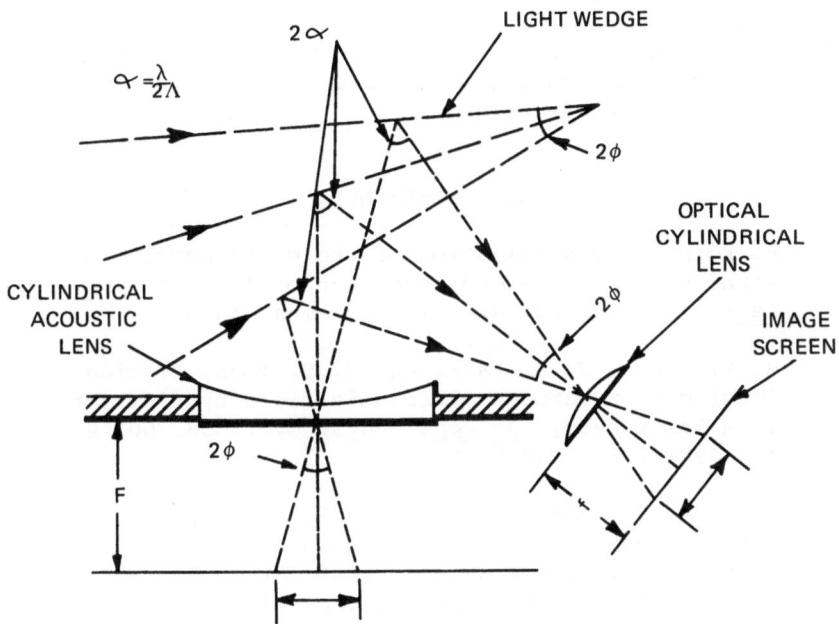

Fig. 14   Diagram showing how field of view is limited by
           the light wedge in a system containing a
           cylindrical acoustic lens.

9.7 FINAL GENERAL REMARKS

As far as practical applications are concerned, Bragg-diffraction imaging may turn out to be more suited for industrial examination of manufactured parts than for any other application. Certainly its suitability in medical diagnosis is questionable (along with all other systems using laser-beam readout) in view of its lack of sensitivity. Whenever good sensitivity is needed, systems using piezoelectric readout will apparently have the advantage. However, in applications other than medical diagnosis, where higher-frequency operation is permissible, Bragg-diffraction imaging may well be quite competitive, particularly if phase retention is important.

9.8 REFERENCES

[1] A. Korpel, "Visualization of the cross-section of a sound beam by Bragg-diffraction of light," Appl. Phys. Lett., Vol. 9, pp. 425-427, December 1966.

[2] H. V. Vance, J. K. Parks and C. S. Tsai, "Optical imaging of a complex ultrasonic field by diffraction of a laser beam," J. Appl. Phys., Vol. 38, No. 4, pp. 1981-1983, March 1967.

[3] G. Wade, J. Landry and A. A. deSouza, "Acoustic transparencies for optical imaging and ultrasonic diffraction," paper at the First International Symposium on Acoustical Holography, December 1967.

[4] M. G. Cohen and E. I. Gordon, "Acoustic beam probing using optical techniques," Bell Syst. Tech. J., Vol. 44, pp. 693-721, April 1965.

[5] J. L. Kreuzer, "Acoustic Bragg imaging with an optical point source," in Acoustical Holography, Vol. 4, G. Wade, Ed. New York: Plenum, 1972, pp. 583-597.

[6] A. Korpel, "Astigmatic imaging properties of Bragg diffraction," J. Acoustical Soc. Am., Vol. 49, pp. 1059-1061, March 1971.

[7] R. A. Smith, G. Wade, J. Powers, and J. Landry, "Studies of resolution in a Bragg-imaging system," J. Acoustical Soc. Am., Vol. 49, pp. 1062-1068, March 1971.

[8] J. Powers, R. Smith and G. Wade, "Phase aberrations in Bragg imaging," in Acoustical Holography, Vol. 3, A. F. Metherell, Ed. New York: Plenum, 1971, pp. 71-91.

[9] L. Brillouin, "Diffusion de la lumiere et des rayons X par un corps transparent homogene influence de l'agitation thermique," Ann. de Physique, Vol. 17, pp. 88-122, 1922.

[10] J. W. Goodman, Introduction to Fourier Optics, San Francisco: McGraw-Hill, 1968, p. 45.

[11] A. Korpel, "Optical Imaging of Ultrasonic Fields by Acoustic Bragg Diffraction," Doctoral Dissertation, University of Delft, Netherlands, 1969.

[12] J. P. Powers, "Spatial Filtering Considerations in Bragg-diffraction Imaging," in Acoustical Holography, Vol. 4, G. Wade, Ed. New York: Plenum, 1972, pp. 533-567.

[13] D. C. Winter, "Noise reduction in acoustal-optic (Bragg) imaging systems by holographic recording," Appl. Phys. Letters, Vol. 22, No. 4, pp. 151-152, Febr. 15, 1973.

[14] S. C. Pei and G. Wade, "Noise analysis of holographic detection of Bragg-diffraction imaging," Digest of Papers for Int'l. Optical Computing Conf., 1975, pp. 124-128, April, 1975.

[15] K. Wang and G. Wade, "Comparison of ideal performance of some real-time acoustic imaging systems," J. Acoust. Soc. Amer., Vol. 56, No. 3, pp. 922-928, Sept. 1974.

[16] J. W. Goodman, "Temporal filtering properties of holograms," Appl. Optics, Vol. 6, pp. 857-859, May 1967.

[17] A. Papoulis, Systems and Transforms with Applications in Optics, New York: McGraw-Hill, 1968, p. 75.

[18]   R. Smith and G. Wade, "Noise characteristics of Bragg
       imaging," Acoustical Holography, Vol. 3, Ed.
       A. F. Metherell, New York: Plenum, 1971, pp. 93-128.

[19]   K. Wang, Threshold Contrast for Various Acoustic
       Imaging Systems, Unpublished Ph.D. Dissertation,
       University of California, Santa Barbara, 1972.

[20]   G. Wade, A. Coello-Vera, L. Schlussler, and S. C. Pei,
       "Acoustic lenses and low-velocity fluids for
       improving Bragg-diffraction images," accepted for
       publication in Acoustical Holography, Vol. 6,
       N. Booth, Ed. New York: Plenum, 1975.

Chapter 10

IMAGING WITH DYNAMIC-RIPPLE DIFFRACTION

Lawrence W. Kessler

Sonoscan, Inc.
752 Foster Avenue
Bensenville, Illinois 60106

Introduction

Most methods of point-by-point scanning of acoustic
fields employ relatively slow mechanical methods to measure
amplitude and phase in the receptor plane. Optical methods
of sound field detection have been discussed in a very
general way in Chapter 3, and it was pointed out that point-
by-point optical scanning has several advantages over direct
optical conversion methods. In particular, since the sound
field parameters are ultimately translated into electrical
signals, electronic filtering may be employed to reject
unwanted background. Furthermore, the choice between conventional imaging and holographic imaging is made by changing from linear detection to phase detection, respectively.
The commercial availability of lasers and laser beam scanning devices permit the construction of acoustic imaging
systems to be relatively straightforward, and with the appropriate devices and TV compatible circuitry, real time
imaging is possible.

The dynamic-ripple effect that occurs when a sound beam
strikes an interface is similar to the periodic deformations
that are produced by surface acoustic wave energy. The
method for detecting the ripples was originally developed
for visualizing surfaces waves[1] and then adapted for rapid
sampling of acoustic holograms[2]. It relies upon the periodic angular modulation which is imparted to a reflected
focused laser beam by the surface tilting. Compared with
the stabilized interferometric technique[3] which is still
somewhat sensitive to path length variations, this technique

appears to be simpler from a practical standpoint, less sensitive to static irregularities of the surface height and less demanding on electrical bandwidth for real time operation with good sensitivity.

Application of this technique to acoustic imaging and holography over the frequency range of 1-500 MHz has been demonstrated [4-8]. At the low frequencies, the imaging systems are called "acoustic cameras" and above about 100 MHz "acoustic microscopes". The ultimate limit on permissible frequency of operation occurs when the wavelength of sound and the wavelength of the laser light become comparable in dimension.

Principle of Operation-Interface Response Characteristics

A plane wave of sound crossing, or being reflected from a non-rigid boundary distorts the interface as shown schematically in Figure 1. The boundary displacement is dynamic and the fluctuation rate is the acoustic frequency. The phase front of the "displacement" propagates to the right at velocity $c_0/\sin\theta$ as discussed in Chapter 3.

In practice the boundary separates water on the bottom and a plastic faceplate on top. Plastic is usually chosen

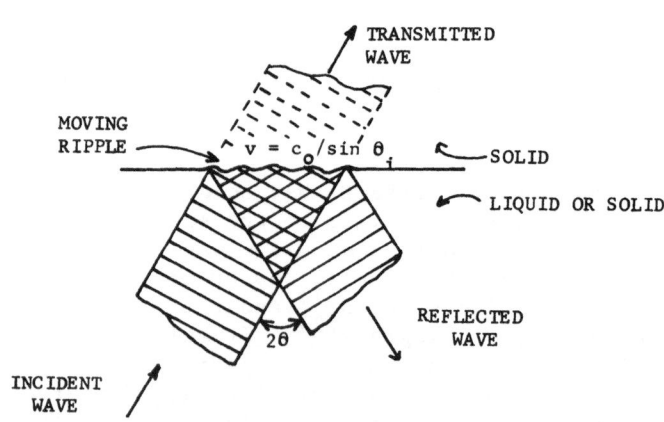

FIGURE 1

since, compared to other solids, its acoustic impedance is
reasonably close to that of the water. It is also quite
lossy acoustically, so that, if desired, a semi-infinite
half-space may be simulated, thereby avoiding resonance
effects. The plastic, which is optically transparent is
metallized at the interface to enhance optical reflectivity.

In an imaging mode, sound scattered from an object in
the liquid is incident upon the interface. Each place wave
in the angular spectrum of the scattered sound field causes
its own characteristic ripple pattern. Frozen in time, the
composite pattern constitutes a hologram of the sound field
recorded with a fictitious reference beam incident normal
to the interface.

The interaction of the "laser beam probe" with a rippled surface depends strongly upon the efective diameter
of the focused laser beam spot, d. In addition, the angular
response characteristic of the surface itself is a function
of the velocity of sound difference between the liquid and
the solid. Thus, good sensitivity and numerical aperture
(NA) of the system depends upon proper consideration of
these design factors. As indicated in Chapter 3, the angular modulation of the reflected laser beam is detected by
a knife edge technique, a simplified schematic of which is
shown in Figure 2. The surface to be examined lies in the

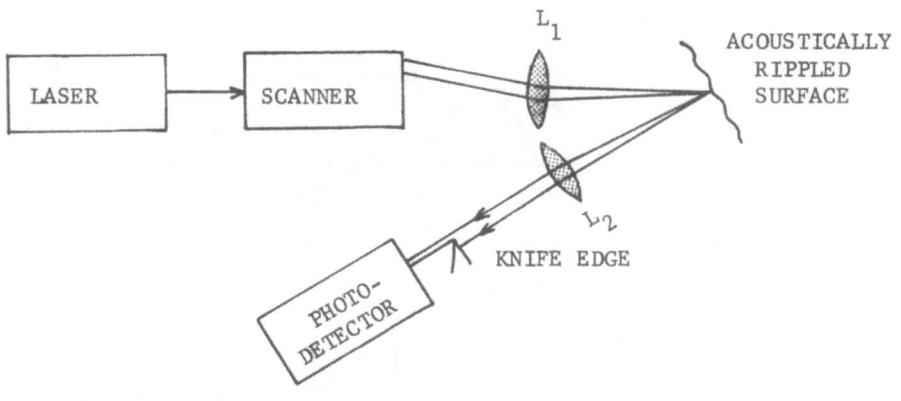

FIGURE 2

Fourier transform plane of the laser beam scanner (i.e.,
the focal plane of lens L_1), thus bringing the probing
light beam to focus on the surface as well as enabling it
to traverse the surface when the laser scan angle is changed.   The knife edge and photodetector lie in the Fourier
transform plane of the surface and, therefore, are in the
image plane of the scanner.   This assures the condition
that the light beam always strikes the knife edge and photodetector in the same place regardless of the laser scan
angle, provided that the surface is not perturbed.

      In Figure 3, the probing light beam is illustrated as
it interacts with the distorted surface.   Here the light
beam diameter is small compared with the spatial periodicity
Λ_s of the surface pattern.   If the wavelength of sound is
Λ, then $\Lambda_s = \Lambda/\sin\theta$.   The relative effect of the light
beam diameter compared with Λ_s on the knife edge response
has been calculated by Whitman and Korpel[9] and is summarized below.   The response function, $G_1(\theta)$, must be considered separately for the three regions of interest which depends upon the ratio d/Λ as follows:

$$G_1(\theta) = (2d/\Lambda)\sin\theta \text{ for } d < \Lambda/2\sin\theta \qquad (1)$$

$$G_1(\theta) = 2 - (2d/\Lambda)\sin\theta \text{ for } \frac{\Lambda}{2\sin\theta} < d < \frac{\Lambda}{\sin\theta} \qquad (2)$$

$$G_1(\theta) = 0 \text{ for } d > \frac{\Lambda}{\sin\theta} \qquad (3)$$

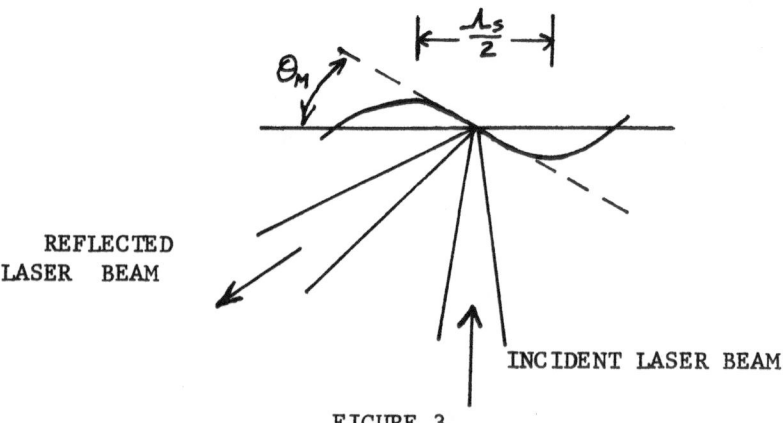

REFLECTED
LASER BEAM

INCIDENT LASER BEAM

FIGURE 3

The physical significance of these formulas is that the response falls to zero for large and small values of d compared with Λ_s. Somewhere inbetween, specifically at d = $\Lambda_s/2$ the response is optimum. As d gets small the angular spread of the light is so great, that a small tilting of the surface is not observable. On the other hand, when d gets large, light is refracted in opposite directions simultanenously thereby cancelling the knife edge output.

In order to determine the response of the total system, the function $G_1(\theta)$ must be multiplied by another function, $G_2(\theta)$, which describes the behavior of the solid-liquid interface as acoustic waves impinge from different angles. Depending upon the material chosen for the faceplate, various overall transfer curves are possible. The ideal response, that is one with no zero between $0°$ and $90°$, would be obtained with a material with acoustic velocity equal to that of water, viz., 1500 m/sec. A typical plastic with which we have been quite successful so far, is lucite, which has a sound velocity of 2670 m/sec., thus giving rise to a critical angle at $34°$. For this material the shear wave velocity is less than the compressional wave velocity in water and, therefore, no second critical angle occurs. However, if a much harder substance such as glass is employed as a faceplate, a second critical angle caused by the shear wave, will restrict the angular aperture even more.

Figure 4 describes the function $G_1(\theta)$ for a water-lucite interface which was computed from a recent theory[11] which takes into account the effects of evanescent waves at the

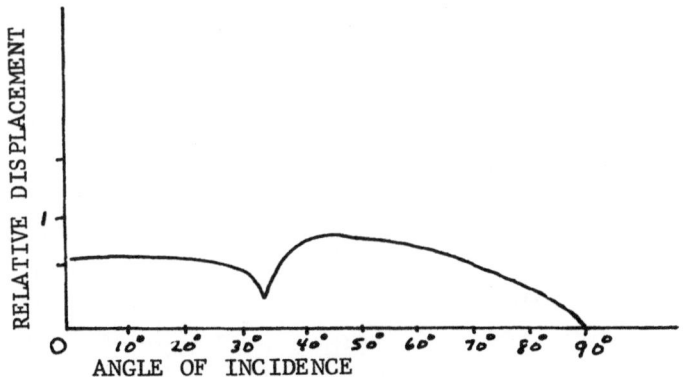

FIGURE 4

interface and acoustic absorption in the plastic material. From these figures, it appears that the angular response can be quite uniform and that a high NA is achieved with this technique.

The output of the photodetector is an electrical signal that is coherent with the sound at each point. If the light beam were stationary, the frequency of this signal would be equal to the sound frequency. However, the scanning motion of the beam causes a Doppler shift, the magnitude of which depends on the relative velocity of the surface ripple and the laser scanning speed. As already mentioned, each plane wave in the angular spectrum of the scattered sound field causes a characteristic ripple pattern, and hence results in a characteristic Doppler component. For rapid scanning, the Doppler shift may be appreciable, for example, a few megahert for real time operation.

Figure 5 illustrates a block diagram of such a system. The image of the object may be focused onto the faceplate or a direct shadow method employed. The system can operate in a holographic mode or direct imaging mode. Furthermore, the overall response of the interface transfer function, G_1 (θ) x $G_2(\theta)$, can be modified electronically to give other modes of operation such as dark field.

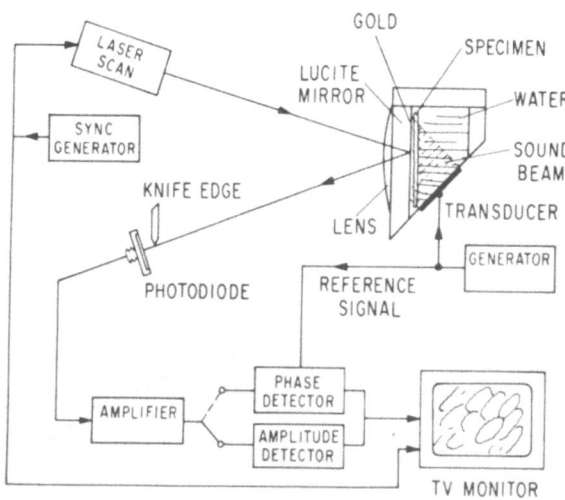

FIGURE 5

In the holographic mode, the composite Doppler signal
is extracted by mixing with the original sound frequency
and fed into a television monitor. This results in a stationary display of the original composite ripple pattern,
i.e., the acoustic hologram. A photograph is then taken of
the television screen and the hologram is reconstructed in
the conventional manner. In the direct, non-holographic
mode, a picture of the sound field may be obtained by replacing the mixer with a detector. The sensitivity of
the knife edge system has been previously derived[2,10]
and it is presented below in terms of the minimum detectable
sound intensity, I_{sm} , for which the signal to noise ratio
is 1. It has been further assumed that the electronic system is shot-noise limited. That is, the laser induced shot-
noise in the photodetector is greater than the electronic
noise of the receiver.

$$I_{sm} (\theta) = \frac{1}{G_1^2 (\theta) \; G_2^2 (\theta)} \; \frac{f^2 \; Z \; \lambda^2 h \nu}{2HP_\ell \; t} \qquad (4)$$

Here, f is the acoustic frequency, Z is the acoustic impedance of water, λ is the optical wavelength, $h\nu$ is the pho-
ton energy, P_ℓ is the laser power, t is the sampling time
(t = 1/2 x Bandwidth), and H is the quantum efficiency of
the photodetector.

Application to Practical Systems

The dynamic ripple diffraction effect has been employed over a wide range of acoustic frequencies. It has been
found that this technique is particularly attractive at
higher frequencies, i.e., the microscopy regime. Figure 6
is a schematic diagram of a commercial apparatus for real
time acoustic microscopy [12] and Figure 7 shows an actual
instrument for use in biomedical and material studies. The
function of the stage is to couple acoustic energy to a
specimen at the desired angle of illumination and to mechanically support it parallel to and in close proximity to
the mirror. The simplest arrangement which like Figure 6
employs a water path coupling between the transducer and
coverslip is not practical because of high acoustic absorption losses in the water. Instead, an optically transparent
and acoustically transparent glass is employed. Although
there are several stage configurations which are suited to
particular applications, in this arrangement, intended for

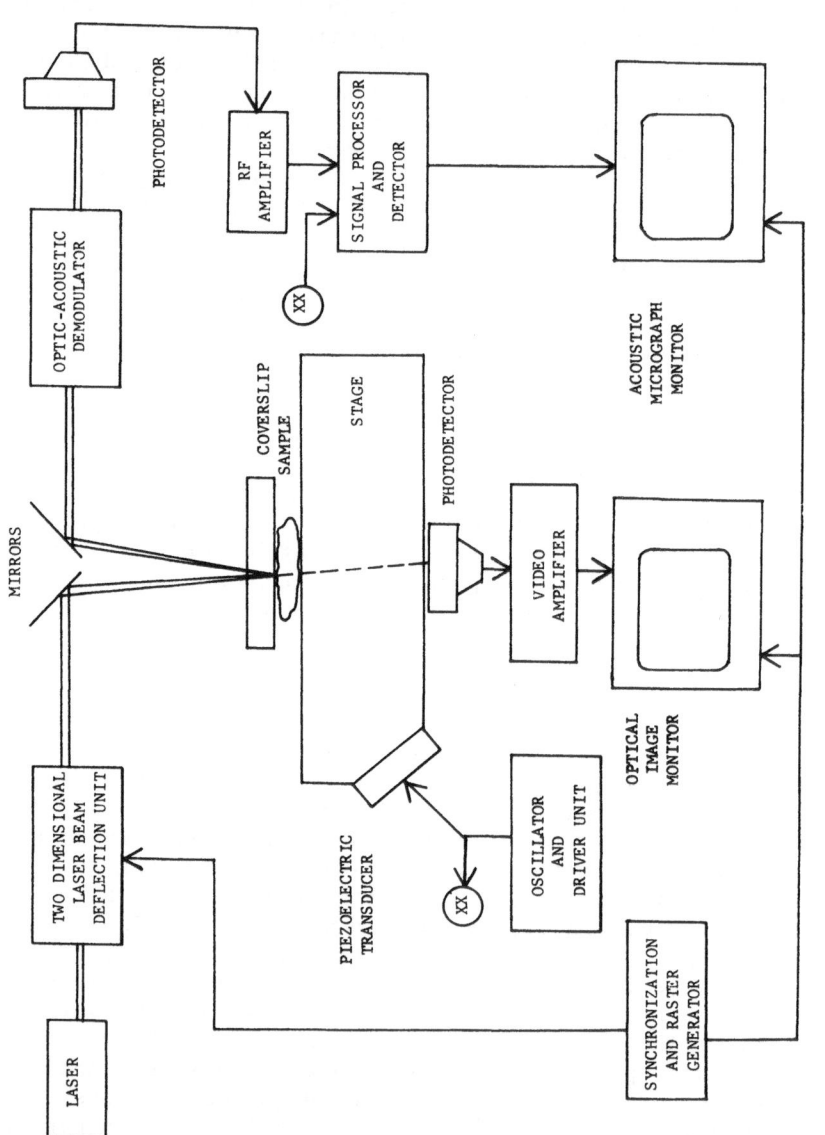

FIGURE 6

FIGURE 7

biological applications, the sound beam is refracted from
an angle of 45° in the solid, to 10° in the sample due to
the large ratio of acoustic velocities. The transmitted
sound intensity is about 10 db below that of the incident
beam and the remaining sound is scattered and absorbed within the block. A plastic mirror which acts as a coverslip
and has a semi-transparent coating is placed directly above
the specimen, sandwiching it between the mirror and quartz
substrate. As a side benefit, a fraction of the probing
light beam is transmitted through the sandwich to a photodiode which receives optical image information on the
specimen. Thus acoustical and optical images of a microscopic size specimen are simultaneously displayed.

In another configuration, intended for inspection of
metals and ceramics, the sample itself may be optically
polished on one surface and thereby avoid the necessity of
a coverslip. Furthermore, when it is not possible to produce an optical through-transmission image of the sample,
an optical reflection image of the surface can be produced
instead. For odd shaped samples, special configuration
stages can be employed for coupling the acoustic energy
from the source to the sample and then to the receptor plane.

References

1) R. Adler, A. Korpel, and P. Desmares. "An Instrument
    for making Surface Waves Visible". IEEE Trans. SU
    Vol. SU-15, 157, (1968)

2) A. Korpel and P. Desmares "Rapid Sampling of Acoustic
    Holograms by Laser Scanning Techniques", J. Acoust.
    Soc. Amer. 45, 881 (1969).

3) R.S. Mezrich, K.F. Etzold and D.H. Vilkomerson "Ultra-
    sonovision" Proc. 1974 Ultrasonics Symposium. Mil-
    waukee, Wisc. IEEE Cat#74 CHO 896-SU pp 1-4

4) A. Korpel, L.W. Kessler amd P.R. Palermo, "An Acoustic
    Microscope Operating at 100 MHz", Nature, 232, 110-111
    (1971).

5) L.W. Kessler, P.R. Palermo and A. Korpel, "Practical
    High Resolution Acoustic Microscopy", _Acoustical_
    Holography, Vol. 4, Plenum Press, New York (1972), ed.
    by G. Wade, pp. 51-71.

6) L.W. Kessler, P.R. Palermo and A. Korpel "Recent Devel-
    opments with the Scanning Laser Acoustic Microscope"
    in _Acoustic Holography Vol.5_, ed. by P. Green, Plenum
    Press, New York (1974).

7) R.L.Whitman, M. Ahmed and A. Korpel "A Progress Report
    on the Laser Scanning Camera" _Acoustical Holography_
    Vol. 4, Plenum Press, NY (1972) ed. by Wade pp. 11-32

8) R.L. Whitman, A. Korpel and M. Ahmed "Novel Techniques
    for Real Time Depth-Gated Acoustic Image Holography"
    Appl. Phys. Lett 20, 370 (1972).

9) R. L. Whitman and A. Korpel "Probing of Acoustic Sur-
    face Pertubations by Coherent Light", Applied Optics,
    8, 1567 (1969).

10) A. Korpel and L.W. Kessler, "Comparison of Methods of
    Acoustic Microscopy", _Acoustic Holography, Vol. 3_, A.
    F. Metherell (ed.), Plenum Press, NY (1971) Ch. 3, pp.
    23-43

11) M. Ahmed, R. L. Whitman and A. Korpel, "Response of an
    Isotropic Acoustic Imaging Faceplate" IEEE Trans. SU
    SU-20 323 (1973).

12) SONOMICROSCOPE 100 manufactured by Sonscan, Inc.
    Bensenville, Illinois 60106 (312)   766-8795

5. E.N. Beasley, R.D. Falzone and A. Gordon, "Practical High Resolution Acoustic Microscopy", Acoustical Holography, Vol. ..., Plenum Press, New York, (19..).

Chapter 11

IMAGING USING LENSES

C. F. Quate

Stanford University

Stanford, California 94305

INTRODUCTION

There are several acoustic imaging systems which are designed to respond to the energy coming from a point source in the form of a spherically diverging wave. The field pattern of a given object can be thought of as a large number of point sources - each with a different amplitude and phase. Thus, if the imaging system can sequentially receive energy from each of these points on the image we can build up the entire pattern. In this chapter we will describe three such systems in order to give the reader some appreciation of the variety of ideas that have evolved. Each system uses a lens to focus on the point source and the scanning is carried out either by mechanical motion of the object through this focal point or by "electronic scanning" wherein the phase of the incoming signal as collected by an array is carefully adjusted to synthesize a spherical lens.

We have chosen three frequency ranges for these examples for the "size of things" has a strong influence on the particular system that is selected for a given task. With the "electronic lens" the work has been carried out near 2 MHz which is suitable for viewing structure within the human body cavity. A linear array of piezoelectric elements is used. With the proper adjustment of the phase at each of these elements for the received signal a spherical wavefront produces a maximum output signal. A plane wavefront in this system produces a diminished output since adjacent elements in the array tend to cancel

241

each other.   The focal point of this lens can be adjusted both in depth and along a lateral line.   The two-dimensional scan is thus in the form of a "B-scan".

In the second system the spherical wavefront is received by a Fresnel lens fabricated with a piezoelectric ceramic.   The focal length of such a lens can be changed by varying the frequency and thus a variable frequency will produce a linear scan along a line normal to the lens surface.   Mechanical motion is used in the transverse dimension to generate a "B-scan" presentation.   The frequency of operation is 10 MHz and this should be useful in ophthalmology.

The third system incorporates a single surface spherical lens between sapphire and water.   This lens has a fixed focal point and the image is recorded by mechanically scanning the object through this focal point in a raster pattern.   The output is converted into an electrical signal with a piezoelectric transducer and displayed on a TV monitor.   This is perhaps one of the least complicated of the acoustic imaging systems and this simplicity has made it possible to operate the instrument at a frequency of 1000 MHz.   There the beam at the focal point approaches one micron in diameter.   This is sufficient to allow us to probe the microscopic detail on the interior of mammalian cells.   We will begin with a discussion of this instrument.   The major portion of our discussion will be devoted to this system since the images are more numerous and of higher quality than those that have been recorded to date with the low frequency systems.

SCANNING ACOUSTIC MICROSCOPE

Background

The acoustic microscope is an instrument designed to exploit the intrinsic property of ultrasonic waves in liquids namely the short wavelengths available at frequencies near 1 GHz $(10^9$ Hz).   Difficulties arise in contemplating an instrument of this type since it is not obvious how one might visualize the pattern of acoustic energy - a pattern that reproduces the essential features of the specimen that is to be viewed.   This problem does not arise with the optical microscope - but with high

frequency acoustic waves there is nothing equivalent to
either the human eye or photographic film. In the system
as described in Chapter 10 by Kessler a scanning light
beam was used to "read" the pattern of acoustic energy.
In the work to be described here the acoustic beam is
focused by a lens and detected by a piezoelectric film.
Mechanically scanning the specimen through the focused
acoustic beam gives us the total image.

The scanning acoustic microscope[1] has been advanced
to the point where it can be used for imaging mammalian
cells and tissues with a resolution that approaches one
micron. This is sufficient to record significant detail
in cell complexes - both normal and neoplastic. The
acoustic instrument responds to the elastic properties of
the object and it, therefore, provides information which
is distinct from the optical microscope. For example,
the acoustic contrast is large in biological material and
it is not necessary to use staining techniques to bring
out the detailed structure. Differences between the
elastic properties of normal and abnormal cells can be
visualized with the acoustic microscope. With cultured
cells it will give information on the viscosity of the
cytoplasm and this may prove important in the study of
abnormal processes within the cell.

We have been able to show that this instrument can
be used to view microscopic detail on integrated circuits.
This work is not a subject for this series, but it is
interesting and the results have appeared in the
literature.[2]

The potential for acoustic waves as an alternative to
optical waves in microscopy was discussed in early work by
Dunn and Fry[3] at the University of Illinois. A full
description of this work is included as a Chapter[4]
entitled "Ultrasonic Microscope" in Clark's Encyclopedia
of Microscopy. They disclosed - and this had been
pointed out even earlier by Sokolov[5] - that acoustic
waves at 1 GHz in water with a wavelength of 1.5 microns
should in principle permit one to construct an instrument
for viewing microscopic structures with a resolution
comparable to the optical microscope. They recognized
that acoustic micrographs could provide increased contrast
without resorting to chemical staining techniques.

In parallel with this work various laboratories have
conducted studies of the sound absorption in mammalian
tissue and other biological material. The work has been
done with frequencies near 1 MHz.   This is far below the
operating frequencies of the microscope but if done with
care the data can be extrapolated to the high frequencies.
Some of the earliest work was that of Carstensen and
Schwan[6,7] who worked with protein and hemoglobin.   They
found that the attenuation was high in comparison to other
liquids such as water.   They established the frequency
and temperature dependence of the absorption coefficient.
This parallel work formed the basis of our belief that
acoustic micrographs would exhibit excellent contrast -
a belief that has been verified by the acoustic images
themselves.

Subsequent to this Kessler[8] studied the attenuating
properties of sound in mammalian tissue with sections from
both rat and human kidney and found that it too was much
higher than water.   Lees and Barber[9] have worked with the
elastic properties of the materials that comprise the
teeth - enamel and dentyne.   In a beautiful demonstration
of the acoustic attenuation of connective tissue Anderson[10]
compares a region of blood clotting with that of normal
blood.

After these initial concepts were established there
was little progress.   This lack of progress toward the
realization of an actual instrument for viewing can be
attributed to the technology associated with acoustic
waves at high frequencies.   At that time it was not
advanced.   It was necessary to await the development of
transducers which could serve to efficiently convert
electromagnetic energy into acoustic energy.   It was also
necessary to await for creative suggestions as to how the
acoustic patterns could be rendered visible.   This
occurred in the late sixties at three different locations -
Zenith Research Laboratory in Chicago, Stanford University
in Palo Alto, and University College London, England.
The groups with Korpel and Kessler[11] at Zenith, with Auld[12]
at Stanford, and with Ash[13] at University College London,
used a variety of techniques to read the acoustic image
but they were all based on some form of scanning with a
focused laser beam.   Cunningham and Quate[14] at Stanford

developed a non-scanning system that utilized the acoustic
radiation pressure on 1 micron latex spheres as immersed
in liquid.   The radiation pressure was sufficient to
condense the latex spheres into a pattern that reproduced -
on a one-to-one scale - the acoustic pattern.   The image
as imprinted on the latex spheres was viewed with a
conventional optical microscope.   The work on these
earlier versions has been covered by a review as written
by Mueller.[15]

At the beginning of 1973 the work had been advanced
to the point where it was possible to record images with
a resolution of 10-20 microns.   This has been treated in
more recent review articles by Kessler[16] and by Korpel.[17]
A later account of this work is given in Chapter 10.

By mid-1973 both the Stanford group[18] and the Zenith
Group[19] had improved the resolution to 5 microns although
most of the recorded micrographs had a resolution in
excess of 10 microns.

These systems all suffered from two disadvantages
(1) since they were dependent upon some form of light in
the final readout of the image - the ultimate resolution
must always be less than the optical microscope and (2) the
required level of acoustic power was rather excessive
(1 mw/cm^2).   Furthermore, the required intensity increased
when the acoustic frequency was raised to improve the
resolution.

For these reasons Quate and Lemons[20,21] turned to a
system which used piezoelectric films for generation and
detection of the acoustic waves.   They combined this with
mechanical scanning in order to record the entire image.
In this system the acoustic intensity is reduced to a
minimum because of the high efficiency of the piezoelectric
detector.   Optical waves are not used in any part of the
acoustic system and the wavelength of light does not,
therefore, influence the ultimate resolution that might be
attainable.   The resolution of the acoustic microscope
now stands at a value that is close but still not equal to
the optical instrument.   This is somewhat ironic since
the physical principles which establish the resolution in
the two cases are entirely different.

Outline of System

The primary problem that has to be faced with any acoustic imaging device is the absence of a medium equivalent to the emulsion of a photographic film. One must devise a different method for recording the image. In our instrument we have adopted a scanning technique wherein we measure the acoustic absorption point by point across the specimen. We present this information as a modulation of intensity of the electron beam of a TV monitor.

The system can be visualized with the aid of the sketch of Fig. 1.   In this system we begin on the left with a cable (1) carrying electromagnetic energy at a selected frequency.   At (2) we use a piezoelectric film to convert the electrical signal to an acoustic plane wave which traverses the crystal (3) and impinges on a lens at the crystal liquid interface (4).   The lens focuses the acoustic energy to a fine waist at point (5) and after that the acoustic beam diverges until it encounters the second lens at point (6).   There it is refocused into a plane wave and it traverses the crystal (7) to the piezoelectric film at point (8).   There it is again converted to an electrical signal.   It is this electrical signal that is used to modulate the intensity of the TV monitor used for final display of the image.   This image is obtained by inserting the object into the liquid cell at the waist of the acoustic beam profile - point (5) of Fig. 1.

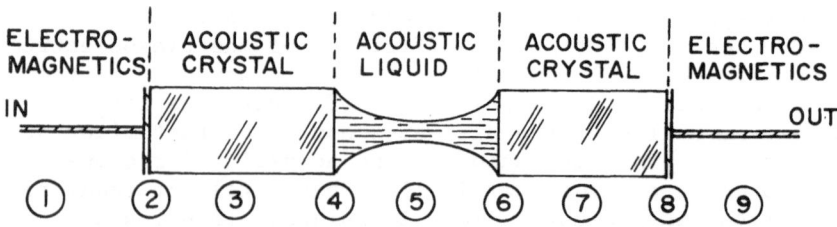

FIG. 1 Sketch of system for acoustic imaging.

The diameter of the beam at this point is less than
the acoustic wavelength in the liquid and it is this
parameter that determines the resolution of the image.
The object is moved through this waist and the variation
in transmission through the acoustic cell is recorded as
variations in intensity on the TV monitor.

The first illustration of what is possible with this
system is shown in Fig. 2.   There we compare the acoustic
and optical image of a metallic grid.   The finite size
of the acoustic beam at the waist (3μ) gives rise to the
slight rounding near the corners but other than this the
image is accurate.   In this example the output signal
changes by something greater than   35 dB   and the transmission through the liquid between the grid wires is
compared to the black regions when the acoustic beam is
blocked by the grid.

As a first example of acoustic images of biological
material we present the photos of Fig. 3.   In this figure
we compare the acoustic and optical images of onion cells.
We know from this that the cell walls absorb or scatter a
great deal of energy as compared to the cellular space

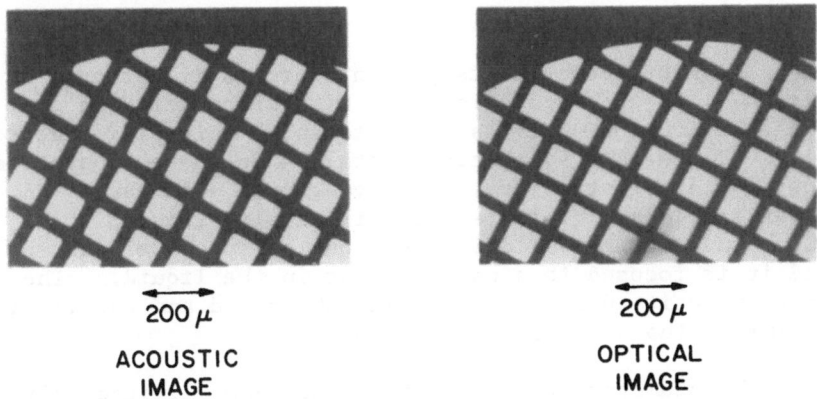

200 μ   200 μ

ACOUSTIC   OPTICAL
IMAGE   IMAGE

FIG. 2 Comparison of the acoustic and optical images of a
200 mesh copper electron microscope grid.

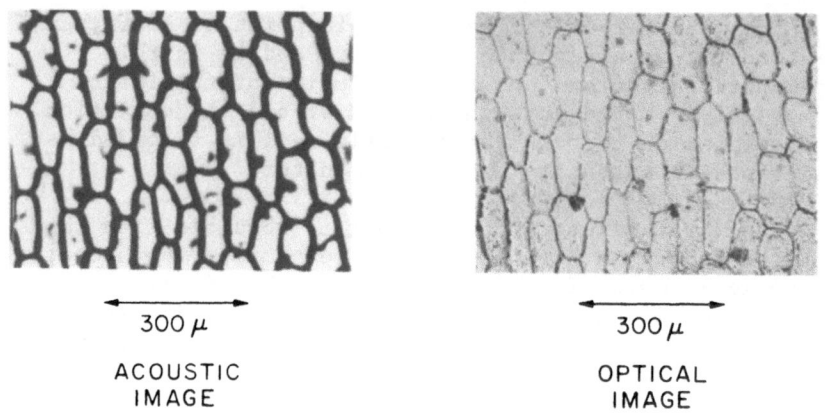

300 μ   300 μ

ACOUSTIC   OPTICAL
IMAGE   IMAGE

FIG. 3 Comparison of the acoustic and optical images of
       onion cells.

between the cell walls.   This gives rise to a contrast
for biological material which is much larger for acoustic
radiation than it is for optical radiation.   This is an
important advantage and we shall return to it in a later
section.

Present Form of the Instrument

The microscope in its present form is illustrated in
Fig. 4.   It uses acoustic waves with frequencies in the
range of 500-1000 MHz as generated with thin piezoelectric
films placed on a synthetic crystal.   On the end opposite
to the piezoelectric film we have ground a concave
spherical lens.   This lens is filled with liquid.   The
plane wave in the solid impinges on the spherical surface
and it is focused to a narrow waist in the liquid.   The
large velocity difference between the solid and the liquid
minimizes the spherical aberration.

The diameter at the waist of the beam, limited only
by diffraction, is a fraction of the acoustic wavelength.
We have a second crystal of similar form symmetrically
placed.   It serves to collect the acoustic beam which
diverges from the waist.   The object to be examined is
mounted on a thin mylar membrane and this in turn is

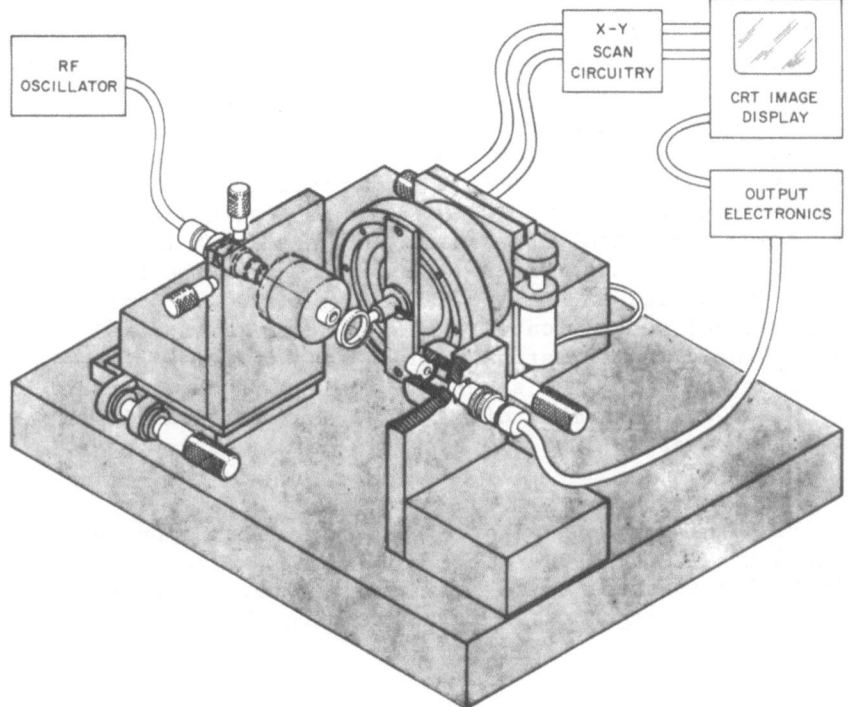

FIG. 4 Generalized diagram of the scanning acoustic
       microscope.

scanned mechanically through the waist of the beam in a
plane perpendicular to the axis.   The acoustic intensity
at the output as a function of time is displayed on a TV
monitor and the raster of this monitor is synchronized
with the mechanical scanning unit to obtain the image
display.

     The resolution of an acoustical microscope is deter-
mined by the quality of its lenses and the wavelength in
medium surrounding the object.   In a scanning type of
microscope this means that one wishes to have lenses with
the largest possible numerical aperture and with the least

spherical aberration. One will use the shortest practicable
wavelength. These requirements can be met by immersing
the object in water which has one of the lowest attenuations
known to occur in those liquids with small acoustic
velocities. The acoustic lens consists of a single
spherical surface dividing two media with differing
propagation velocities. It is interesting to note that
a system such as this would not work for optical waves -
the reason, spherical aberration. The smaller the radius
of curvature and the larger the ratio between the propagation velocities of the two adjoining media the smaller
the spherical aberration will be. With water as one
medium, the present instrument uses sapphire crystals as
the other medium. The spherical lens has a diameter of
0.2 mm and an f number of 0.75. A simplified diagram
showing the assembly of the essential elements is given in
Fig. 5.

With this arrangement the spherical aberration is
reduced to the point where the beam size is determined by
diffraction effects rather than aberrations. This is
illustrated in Fig. 6. There we compare the optical rays

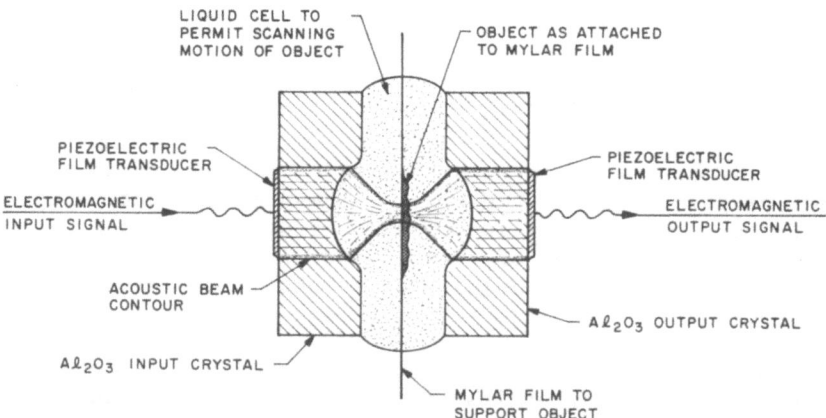

FIG. 5 Schematic showing acoustic lens configuration.

of a single spherical lens consisting of a flint glass air
interface (velocity ratio of 1.52) to those of the
acoustical rays in a similar lens with a sapphire-water
interface (velocity ratio of 7.45). We see that the
aberration in the optical lens is enormous and it limits
the minimum diameter to 25-30 microns. With the acoustic
lens the aberration is very small and the beam diameter as
limited by this effect is less than one micron. This is
smaller than the diffraction limited diameter. It is
this essential feature that allows us to achieve the
resolution of the photos in the section entitled "Present
Work on Mammalian Cells and Tissue Sections".

We shall first describe the transmission mode of
operation as illustrated in the configuration of Figs. 4
and 5. A plane acoustic wave, generated by a planar zinc-
oxide transducer impinges on the acoustic lens and is
focused to a spot which we estimate to have a diameter
below two microns. Part of this radiation is reflected,

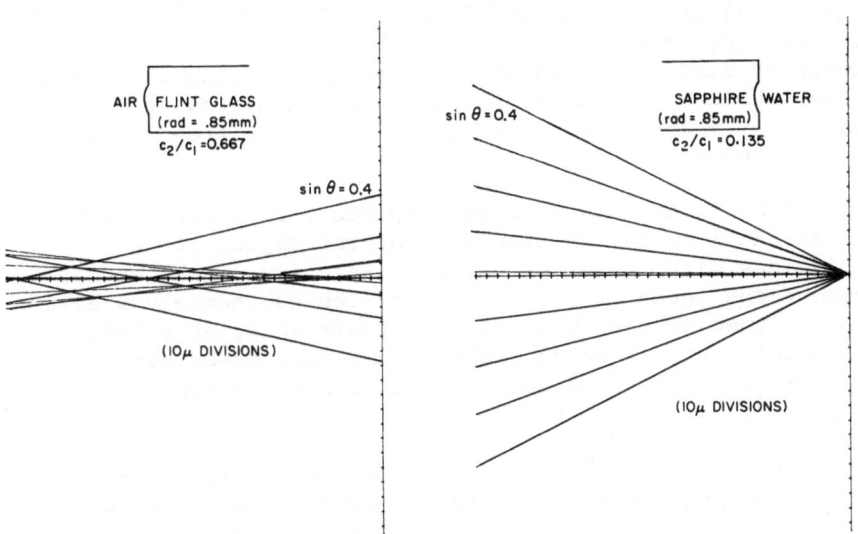

FIG. 6 Ray tracing comparison of the performance of a
single surface lens in a light optical system (left)
and an acoustic system (right). The paraxial focus
lies at the ordinate.

or scattered, by the object placed in the focal plane,
part is absorbed by the object, and part is transmitted.
After passing through the object the radiation will expand
with approximately spherical wavefronts until it impinges
on the second, or collimating, lens, which makes the wavefronts planar so that they now propagate in the form of a
plane wave through the second sapphire crystal and impinge
on the second transducer.   This transducer consists of a
thin layer of zinc-oxide between metal electrodes.
ZnO is piezoelectric, so that the oscillatory strain
associated with the acoustic wave will be converted into
an electric field which, after amplification and rectification, is used to modulate the brightness of a C.R.O.
display.

In order to obtain the usual two-dimensional image of
the object, the object itself is moved across the focus of
the acoustic radiation mechanically in a television-like
scan, and the electron beam of the C.R.O. display is moved
in synchronism with it.   The motion in the x-direction is
sinusoidal in time and is brought about by fastening the
object holder - a thin and light metal ring - to the
movement of an electromagnet.   The y-motion is brought
about by displacing the whole assembly of loudspeaker and
object holder in a vertical direction in a slow and uniform
manner by means of displacing a hydraulic piston driven by
a small pump.

A micrometer is used to displace the object assembly
in an axial or z-direction so that the object plane
coincides with the narrowest cross-section of the acoustic
beam - the focus or "waist".   The lenses themselves have
to be lined up carefully so that their foci, or waists,
coincide; this is accomplished by micrometer-driven motions
in the x, y, and z, directions until maximum signal is
obtained.

At the acoustic frequencies normally used in our
instrument the attenuation in the liquid medium, water, is
high enough so that stray acoustic radiation due to
multiple reflections does not cause degradation in the
image, provided care is taken to shield the electrical
input and output connections from each other.

It is the basic simplicity of the construction of this
kind of scanning microscope which makes it possible to

approach the theoretical limit in resolution; other more
elegant but more complicated schemes have been built though
their performances are still far from the theoretically
possible ones.   Mechanical scanning seems to be a small
price to pay for such a large benefit.

We will now describe the reflection mode of operating
the scanning acoustic microscope as sketched in Fig. 7.
In this mode the "signal" we wish to display consists of
the acoustic radiation reflected from different portions of
the object and/or the phase shifts suffered by the
reflected radiation as a consequence of the deviations
from a plane of the various portions of the object.

The concentrating, or focusing, lens is used also as
a receiving, or collimating, lens, and the input transducer is used as the output transducer.   One can then
dispense with the second lens.   If we wish to record the
amplitude distribution of the object, this can be achieved
in two fundamentally different ways.   In the reflection
mode with continuous radiation one separates the large,
but constant, electrical input power from the feeble, but

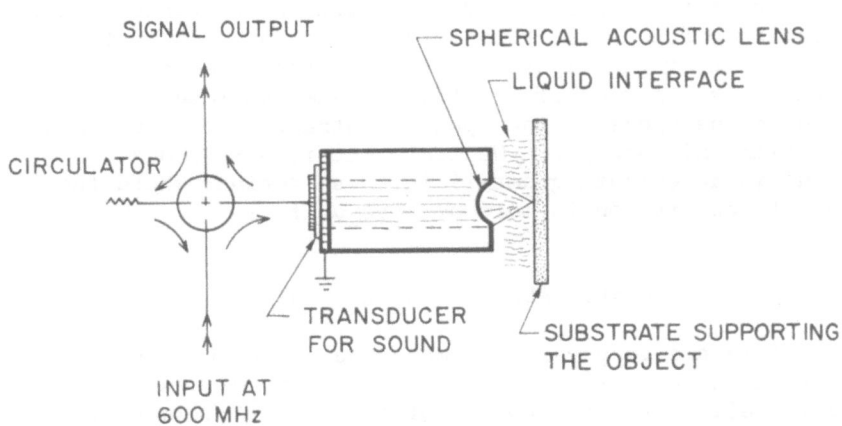

FIG. 7 Diagrammatic scheme for the Reflection Mode.
The substrate is scanned mechanically to obtain
the image.

varying, output power by means of a circulator; this has
been done successfully as shown in Fig. 7.   If the ratio
between input and output powers is too large, so large
that circulator discrimination against leakage is
inadequate, one can operate in the pulsed reflection mode
where one relies on the fact that signal pulses and input
pulses need not arrive at the circulator at the same time.
It is, therefore, possible to "gate" the signal pulses
and suppress the leakage from the input pulses.

    If one wishes to obtain a signal related to the
deviations from planarity of the object, one can super-
impose the reflected signal on a larger and steady "local
oscillator" signal before rectification.   One then
obtains an amplitude distribution which bears a direct
relation to the phase shifts suffered by the reflected
wave.   An example of what can be achieved with this mode
of operation is given in Fig. 8 where we show the image of
a metal grid as immersed in water.   Irregularities along
the edge of the grid bars show up quite clearly.   Other
examples of this technique as used with integrated
circuits have appeared in the literature.[2]

    A third mode of operation - which we term phase
contrast - has been demonstrated in the laboratory.
In this method we compare the phase of the signal as
transmitted through the microscope with a reference signal
at the same frequency.   In this way we can display the
phase shift of the signal relative to the reference
signal as a function of position on the specimen.   This
is quite analogous to the "phase contrast" method as used
in optical microscopy.   We feel that it will prove
useful in accentuating detail in those regions where the
acoustic absorption is more or less uniform.

                    Acoustic Absorption in Liquids

    If we are to physically move the object through the
waist of the beam it is necessary to form this waist in a
liquid cell.   The absorption of sound in liquids is much
larger than in solids and this property controls the
maximum operating frequency of the acoustic system.   If
we are to improve the resolution and reduce the minimum
detectable spot size we must work with higher and higher
frequencies.   In turn, this means that we must understand

FIG. 8 Acoustic reflection image (600 MHz) of a 500 mesh
nickel grid. The periodicity of the grid is ~ 50 μm.

what it is that contributes to the attenuation of sound
in liquids at high frequencies for with this knowledge
we can search for the liquid with the lowest absorption.
Water is one of the most interesting liquids for it has
reasonable loss characteristics and biological samples
are easily immersed in this fluid without undue damage.
However, the absorption in water is sufficient to limit
our frequency to 1000 MHz.   If we are to exceed the
value we must find a suitable method for reducing the
attenuation of water or we must turn to another liquid.
Before we respond to that path of inquiry we want to
outline some of the general properties of liquid
absorption of acoustic energy.

The basic relation for acoustic attenuation in
liquids can be written in terms of the shear viscosity,
η_s , in the form

$$\alpha = \frac{2\omega^2 \, \eta_s}{3\rho v_s^{\,3}} \tag{1}$$

Here α is the attenuation coefficient for the waves which propagate in the form $e^{j(\omega t - kz)} \, e^{-\alpha z}$. In eq. (1) ρ is the density of the liquid, v_s is the sound velocity and ω is the radial frequency. In simple liquids such as neon, argon and xenon the absorption is accurately described by this expression.[22]

In other liquids the situation is more complex. There can be structural changes such as in water where the structure is made up of two parts - one an open structure which is similar to the structure of ice and second the close packed structure of the liquid itself. The attenuation which results from the pressure induced changes between these two can be accounted for by the addition of a second constant to eq. (1). The term is labeled η_v and it is commonly called the volume viscosity. Further than this we can have energy losses as a result of the cyclic variations in temperature created by the excess pressure at the crest of the acoustic wave.[23]

With the inclusion of these terms the equation for α can be written in the form

$$\alpha = \frac{\omega^2}{2\rho c^3} \left[\frac{4}{3} \eta_s + \eta_v + (k/c_v - k/c_p) \right] \tag{2a}$$

or

$$\alpha f^2 = \frac{2\pi^2}{\rho c^3} \left[\frac{4}{3} \eta_s + \eta_v + k \left(\frac{1}{c_v} - \frac{1}{c_p} \right) \right] \tag{2b}$$

A clear discussion of the derivation of this equation is given by Squires.[24] Here k is the thermal conductivity, c_v is the specific heat at constant volume and c_p is the specific heat at constant pressure. The third term in the brackets on the right is small as compared to η_s and is usually neglected. The second term, η_v, (the volume viscosity) is included since the measured attenuation is often two or three times that predicted with the first term alone. The expression in eq. (2) is

still inadequate for in certain molecular liquids the
absorption can be greatly increased over that given in
eq. (2) through a variety of mechanisms all of which are
reviewed in a recent book "Introduction to Clinical
Ultrasonics" M. J. Blandamer, Academic Press (1973).

Even though the heat conductivity related to the
temperature variation within the liquid does not play an
important role, the periodic temperature variation does
give rise in several cases to the perturbation of molecular
equilibrium states. This perturbation represents an
important contribution to α . It has been extensively
studied for it provides information on the energy levels
within the molecules which make up the liquid. This
absorption arises fundamentally from the time delay that
is required to establish equilibrium between the trans-
lational, rotational and vibrational degrees of freedom
of the molecule. The relaxation due to isomorphic
rotation of the molecule about a given axis occur at
moderate frequencies below 300 MHz whereas the
vibrational modes usually occur at frequencies higher than
1000 MHz. The liquid carbon disulphide is an exception
and we will discuss this below.

It is convenient to rewrite eq. (2b) as the sum of
two terms

$$\alpha/f^2 = B + \frac{A}{1 + (f/f_c)^2} \tag{3}$$

Here B represents the frequency independent term
and A represents the contribution due to a molecular
mode with a characteristic frequency of f_c . At
moderate frequencies the A term arising from rotational
motion of the molecule is dominant since its magnitude is
much larger than the magnitude of the B term.

Ethane derivatives represent a classical example of
rotational molecular motion.[25] The molecule as represented
in Fig. 9 can rotate about the carbon-carbon bond. When
one of the hydrogen atoms is replaced by another atom,
rotation about this axis gives rise to different energy
states which correspond to different orientations of the
chlorine-hydrogen triangle as illustrated in Fig. 10.
The resulting absorption curves are given in Fig. 11.

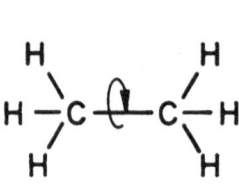

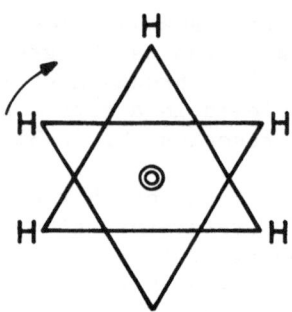

FIG. 9 The ethane molecule. (Lamb).

We see there the high absorption peak at 10.9 MHz and the
leveling out at the higher frequencies.

More interesting from our point of view are the ether's
for the value of the parameter B in those liquids is
25×10^{-17} which is near that of water.   We will keep in
mind that this is the residual value of an absorption at
frequencies far above the characteristic frequency, f_c .
In the ether's f_c is of the order of 200 MHz.

Usually the vibrational modes of the molecules occur
at frequencies higher than 1000 MHz but the liquid, CS_2 ,
is an important exception as mentioned previously.   It
exhibits a vibrational mode with a characteristic frequency
of 78 MHz.[26] The mode is a bending moment within the
molecule itself.

With our work in microscopy we are interested in the
high frequency region.   Carbon disulphide, with its extreme
peak at 78 MHz , is interesting since it exhibits a lower
value of attenuation in a centimeter of length at 3 GHz
than it does at 2 GHz .   More recent work by J. Attal[27]
is given in Fig. 12.   These results were obtained by
measuring the direct acoustic absorption in a liquid cell
of known length.   These results (consistent with those
reported in ref. 26) indicate that we may be able to
operate this instrument above 2000 MHz and thereby record
objects with sub-micron dimensions.   However, CS_2 is a
difficult liquid to work with and our primary interest must

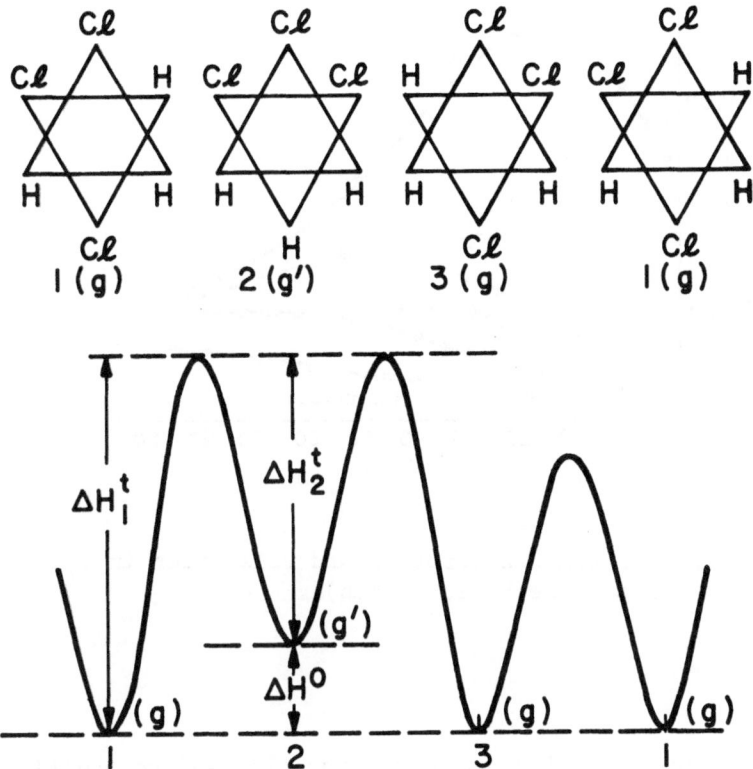

FIG. 10 Rotational isomeric equilibrium in 1,1,2-
trichloroethane. (Lamb).

remain with water.

We want to emphasize some of the important features
of the absorption in this liquid and point out that the
intrinsic absorption can be diminished somewhat with
selected additives.   The ratio, α/f^2 , for the liquid
water is constant over a wide range of frequencies and for
a temperature of $25°C$ it is equal to $24 \times 10^{-17} sec^2 cm^{-1}$.
For a frequency of 1000 MHz this converts to an
attenuation of 2100 dB/cm.   It is the point of reference
that we can use to compare with other liquids.

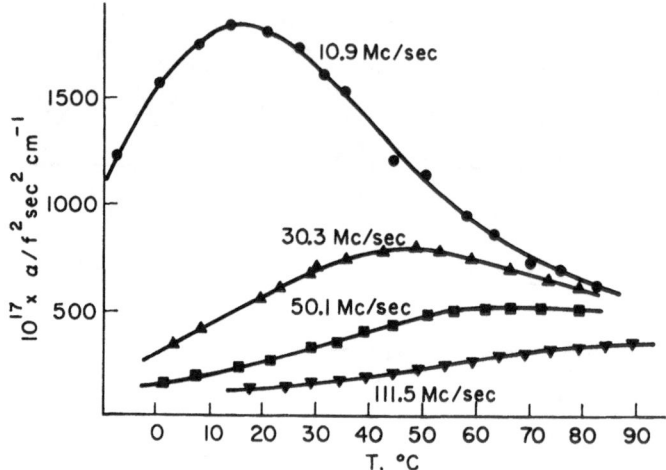

FIG. 11 Ultrasonic absorption and relaxation in 1,1,2-
          trichloroethane.   (Lamb).

        In the model that is conventionally used to explain
the residual absorption the liquid is assumed to consist
of two states - each of a different density.   With
acoustic pressure the molecules will "flow" into the
lattice of closer packing thereby filling up some of the
holes in the liquid.   The rearrangement of structure takes
time and as a result the volume changes are out of phase
with the pressure changes.   This gives rise to the added
absorption.   The theory was worked out by Hall[28] who
proposed that water consisted of a mixture of molecules
partly in an open ice-like structure and partly in a more
closely packed structure.   The relaxation frequency for
the transition between these two states is well above
2000 MHz.

        The absorption coefficient of pure water can be
reduced over the room temperature value by heating - a
procedure that leads to difficulties in an instrument that
is used for viewing living cells and fixed tissue sections.

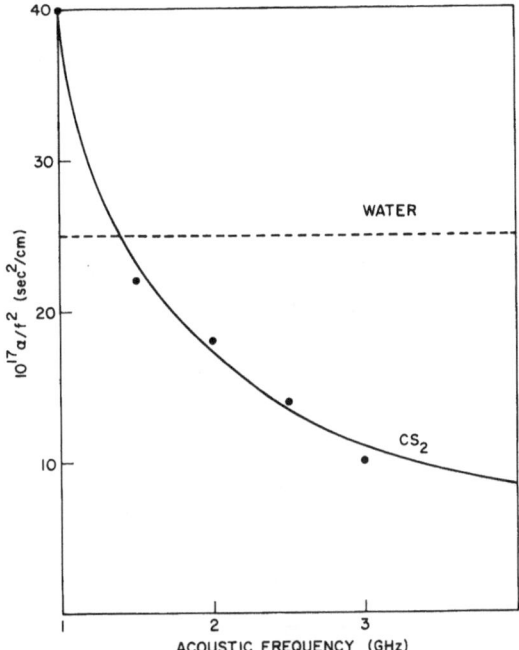

FIG. 12 The absorption of CS$_2$ as compared to that of water - Data is from unpublished work of J. Attal and was obtained with acoustic transmission through a liquid cell.

As an alternative to heating we may want to use liquid mixtures. The work on electrolytes as reviewed by Stuehr and Yeager[29] indicates that there are some electrolytes, when added to water, which serve to lower the value of α/f^2 . The physical mechanism for this lowering is not yet completely understood. But, the effect is clearly evident from the experimental work that has been carried out with NaBr and KBr near 100 MHz. An example is shown in Fig. 13 as taken from the work of Breitschwerdt.[30] Most authors adhere to the hypothesis that the solvent molecules in water increases the rate of conversion between the two structural states in water and thereby brings the volume changes more nearly into phase with the pressure changes.

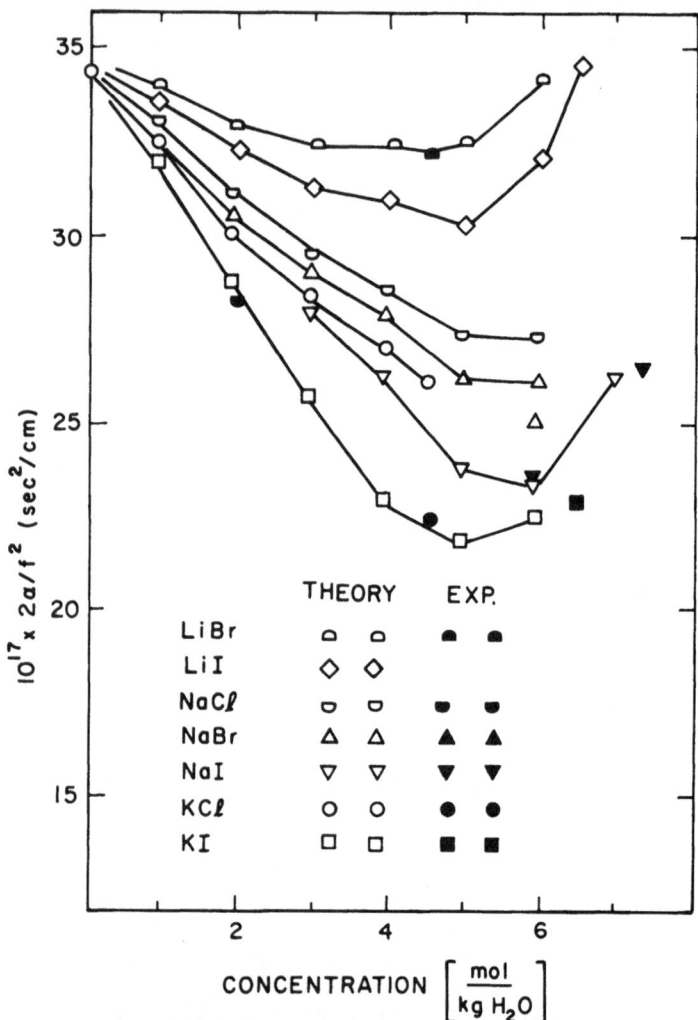

FIG. 13 Comparison between experimental and theoretical
values of the structural absorption in different
ionic solutions. (Breitschwerdt).

Dr. J. Attal in unpublished work has extended these
measurements to higher frequencies. He finds that the
value of α/f^2 is constant over the range of frequencies
from 500 MHz to 4 GHz. The optimum results have been
obtained for concentration of salt near the saturation of

the liquid and these are the results given in Table I.

TABLE I

	Pure water	NaCl	NaBr	NaI	KCl	KI	KCN
10^{17} x $\alpha/f^2(s^2cm^{-1})$	25	20	18	17	16.5	15	13

We should note that there is an increase of the sound velocity of about 15% in most of these mixtures.

It is not clear that these particular mixtures will prove compatible with biological material.   Nonetheless, the work is indicative of what can be done to lower the attenuation in liquid water.   It is the kind of infor- mation that is needed if we are to devise an instrument that will work above 1 GHz and provide a resolution that is more competitive with the optical instrument.

Absorption in Biological Material

The foundation of much of our imaging is based on the absorption of ultrasonic waves in biological material. It is, therefore, important to know in a quantitative manner the values of the attenuation coefficient for the wide variety of materials that are normally encountered. There has not been extensive work in this area but we can discuss a few cases carried out at low frequencies and illustrative of what can be expected.

One of the most interesting examples comes from the published work of Carstensen and Schwan.[1]   They have carefully measured the absorption solutions of mammalian hemoglobin.   This follows earlier work which demonstrated that in blood the sound absorption was primarily due to protein.

A typical plot of the absorption in beef hemoglobin is given in Fig. 14.   The velocity of sound in this material is given in Fig. 15.   The term $\alpha\lambda$ as plotted in Fig. 14 is related to our term α/f^2 and is given by

$$\alpha/f^2 = (\alpha\lambda) \ vf \tag{4}$$

For example, at 2 MHz with the solution of 19.2 gm Hb/100 cc the value of vf is 3.125×10^{11} and the value of α/f^2 is 10^{-14} sec^2/cm. Thus we recognize that this value is two orders of magnitude higher than the absorption in water. This is the one reason for the high contrast that we are able to achieve in our acoustic micrographs.

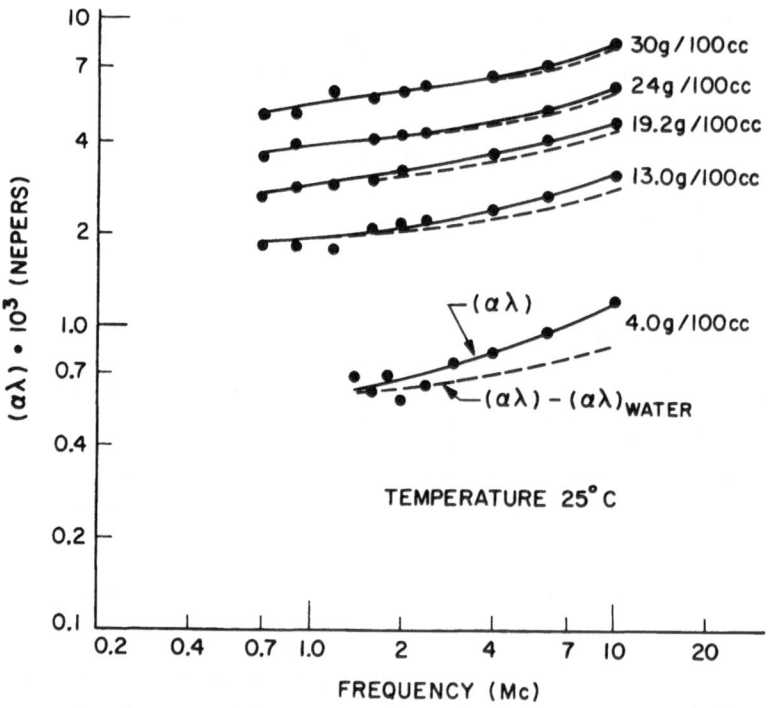

FIG. 14 Absorption of sound in solutions of beef hemoglobin. (Carstensen).

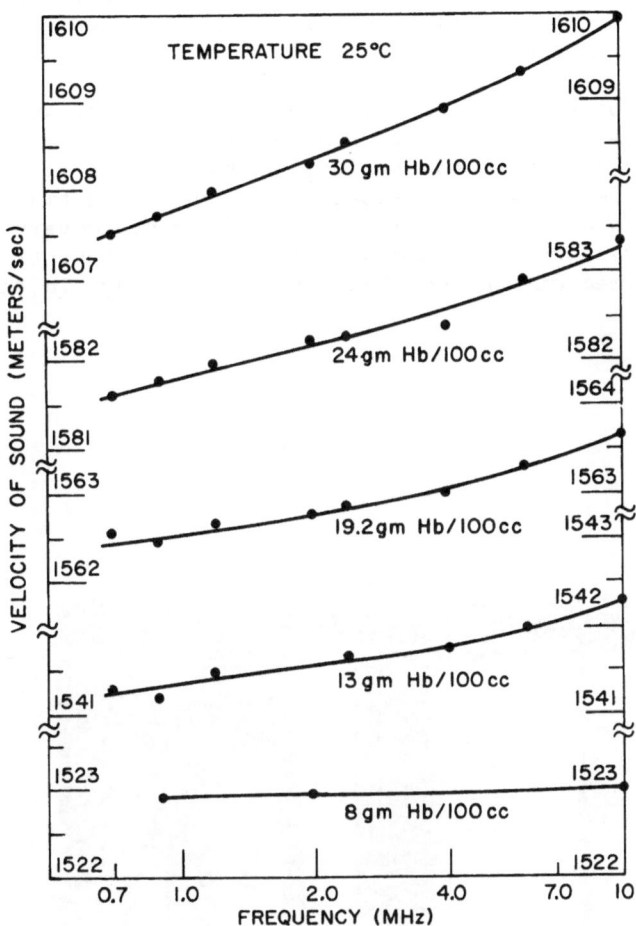

FIG. 15 Dispersion of the velocity of sound in solutions
of beef hemoglobin 25°C. (Carstensen).

This data is valid for the low frequencies as labeled here. One must be careful in extrapolating to the higher frequencies used in the microscope. We don't have the information which would permit us to extrapolate the frequency by a factor of 50. But the α coefficient would still exceed that of water at 1 GHz even with a linear increase with frequency.

In another example of increased attenuation we will turn to the work of Dr. R. E. Anderson[10] at the University of Utah. As mentioned earlier, he did a comparative study of the acoustic attenuation in connective tissue. The results are shown in Fig. 16. There he compares the acoustic response in blood with that of the connective tissue formed with blood clotting. The change in the

FIG. 16 X-ray image on left, ultrasound image on right. (Anderson).

clotting region is clearly evident.   On the left we show
the same material as viewed with x-rays and we see that
with radiation from x-rays the clotting region is not
distinguishable.   Again, these results are for low
frequencies but they are consistent with the work at 1 GHz.
This is not a quantitative result since we don't know the
value of the absorption per unit of length nor the change
in the level of acoustic impedance between the two cases,
but it is a clear indication of the increased attenuation
in connective tissue.

Present Work on Mammalian Cells and Tissue Sections

The acoustic microscope in the form as described in
previous sections has been used to examine a variety of
specimens.   We have accumulated a number of acoustic
images and we are beginning to catalog the response of
various biological materials to acoustic waves.   In this
initial survey we have seen clear distinctions between the
acoustic and the corresponding optical micrographs and we
include a number of these comparative micrographs here.

The samples can be conveniently divided into three
categories.   First, the results as obtained with simple
cell systems containing isolated cells of only a few types
second, living cells grown in culture and third, the tissue
sections which contain a complex collection of cell types.
We have included below a selection of acoustic micrographs
from each of these categories.   The majority of these
images were obtained with an acoustic frequency of
600 MHz $(\lambda = 2.5 \ \mu m)$.   The more recent images, however,
were made at a frequency of 900 MHz $(\lambda = 1.7 \ \mu m)$ with a
resulting resolution which approaches one micron.   In
presenting these images a convention was chosen such that
points on the sample with greater acoustic transmission
are lighter in the corresponding image.   Thus, dark areas
correspond to regions of large acoustic attenuation.

(A) <u>Simple Cell Systems.</u>   As a first illustration,
Fig. 17 shows both acoustic and optical images of a human
bone marrow smear.   The most numerous components in this
smear are the erythrocytes and, again, they show the large
acoustic absorption.   In addition, a number of myeloid
cells can be seen.   These cells are the developmental
elements of the leukocyte family.   In the acoustic

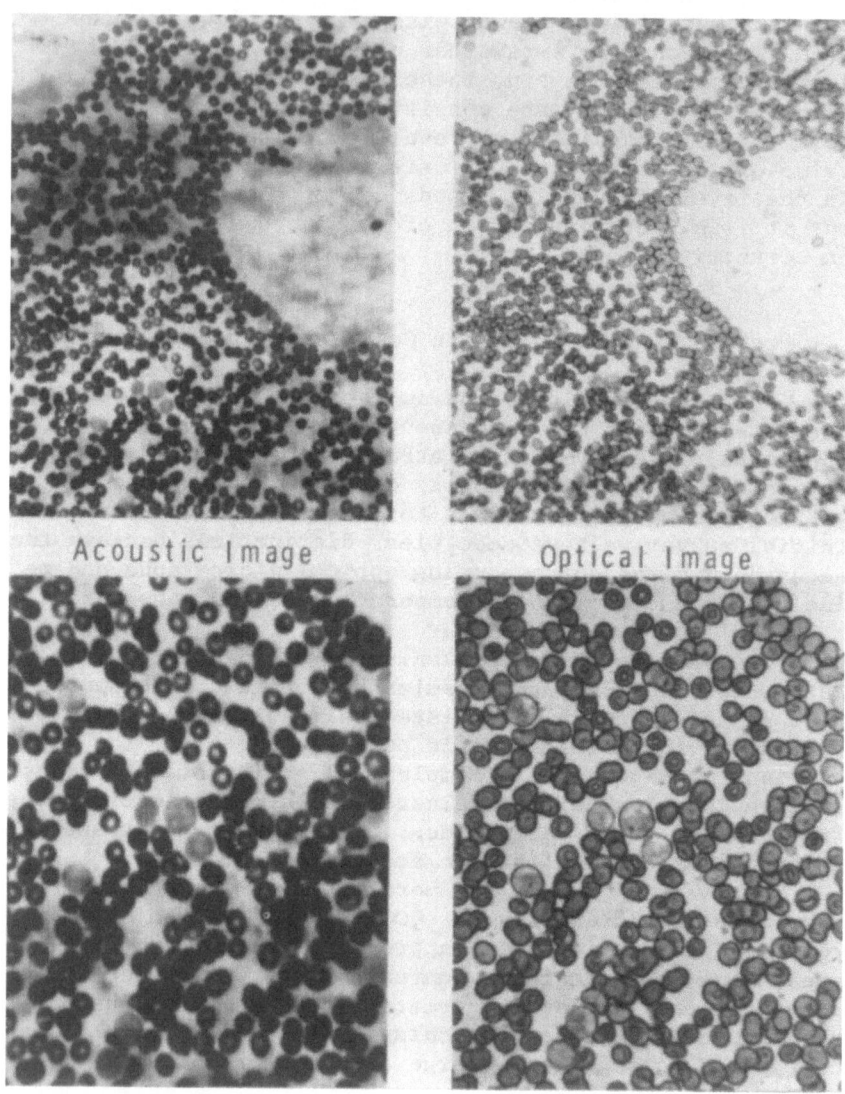

FIG. 17 Comparison of the acoustic (900 MHz) and optical
       images of a human bone marrow smear (stain: hema-
       toxylin-eosin). Magnification 250X at top, 500X
       at bottom.

micrograph they appear larger than the erythrocytes and
they have a markedly lower acoustic attenuation.   In some
instances detail internal to the cell can be seen.   To
bring this detail out in the optical image the sample was
stained with hematoxylin-eosin.   The nuclei stain blue -
a feature which allows them to be easily distinguished from
the cytoplasm.

The contrast that appears in the red blood corpuscle
images is due primarily to the increased absorption of the
hemoglobin within the cell.   This absorption in hemoglobin
has been the subject of previous studies with acoustic
waves at low frequencies.[6,7] These studies show the
absorption to be related both to the viscosity and to the
molecular structure of the material.   The acoustic
microscope is, therefore, sensitive to these parameters,
and can map them on a microscopic scale.   Accordingly, the
acoustic microscope may be useful in studying cells which
exhibit abnormalities in these two properties.

The large inherent differences in the attenuation of
acoustic waves is a powerful feature of the acoustic
microscope that is absent in the optical instrument.

(B) Living Normal Human Diploid Cells.   Cell cultures
represent an important component in the experimental work
on living cells, normal and abnormal.   We feel, for a
number of reasons, that the acoustic microscope can be
applied fruitfully to this area of research.   We have
only begun to explore this field but the acoustic images
as shown in Fig. 18 demonstrate the compatibility of the
acoustic microscope with a living human cell culture.
To date we have demonstrated that cells can be cultured on
the mylar support membrane used in the acoustic microscope.
Moreover, they stick with sufficient tenacity to withstand
the forces that are encountered in a mechanical scanning
system of this type.   In addition, a growth medium can be
substituted for the water between the acoustic lenses in
order to sustain the life of these cells.

The acoustic images of Fig. 18 show a living culture
of normal human diploid lung fibroblast cells.   In this
culture even though the cells are nearly confluent the
extended spindle shape of individual cells can be seen.
The nuclear region of an individual cell is acoustically
more absorbing than are the processes which extend from

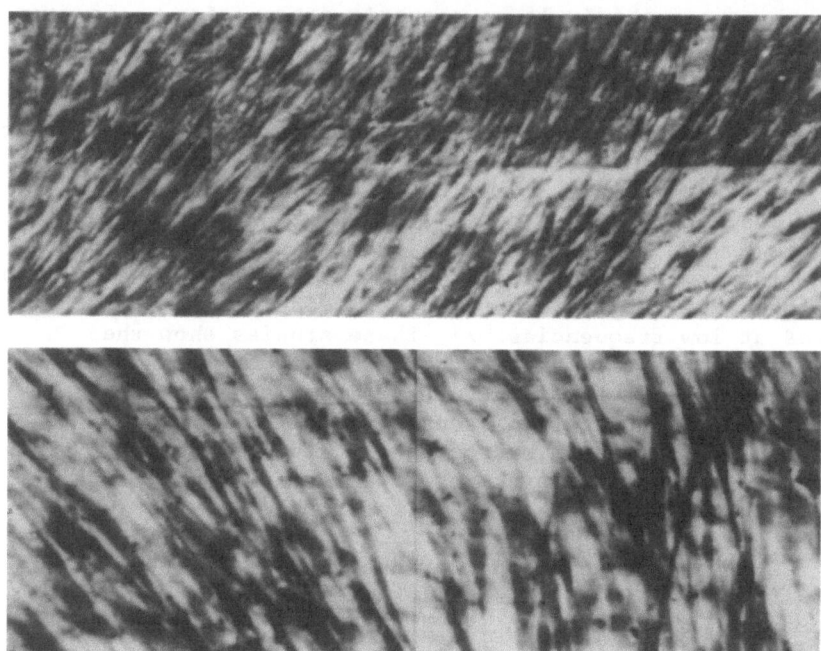

FIG. 18 Acoustic images (900 MHz) of a living culture of
        normal human diploid lung fibroblasts. Magnifi-
        cation ~ 200X at top, 400X at bottom.

each cell.   In some instances small areas of increased
absorption can be seen within the nuclear region.   These
probably correspond to the nucleoli which often occur in
pairs.   The acoustic contrast of these cells is not great.
In an optical microscope these cells are so nearly transparent that phase contrast techniques must be used to
visualize them.   Application of acoustic phase contrast
may well enhance the acoustic contrast of these cells as
well.   We are indebted to the laboratory of Dr. L. Hayflick
for preparing these cell cultures.

    The prime advantage of the acoustic microscope lies
in the different sources of acoustic and optical contrast.
For example, in liquids the acoustic absorption is directly

proportional to the viscosity.   Therefore, changes in
viscosity should appear as changes in the absorption of the
specimen.   There is some early work by Carlson who used
the phenomena of Brownian motion to monitor changes in the
viscosity of the cytoplasm during mitosis.[31]   It was a
rather complex experimental technique but, nevertheless, it
did show a significant and regular change in the viscosity
during the division process.   With the acoustic microscope
it should be possible to make a direct observation of these
changes.

     (C) **Tissue Sections - (1) Normal Tissue.**   In the
next series of acoustic photos we will present the images
of a number of tissue sections - each selected to
illustrate the power of this instrument in viewing
complicated cell systems.   Each of the specimens was cut
from a paraffin block with standard microtome techniques
to a nominal $5 \mu m$ thickness.   We are indebted to the
laboratory of Dr. R. Lawson of the Pathology Department at
Stanford, for preparing these sections.   All of the
tissues were unstained and, therefore, the acoustic
response is typical of what can be expected from the
tissue as altered only by the procedures of fixation.

     In Fig. 19 we present the acoustic image of a section
of human lung tissue.   In this image a number of the
characteristic features of lung tissue are evident.   In
the lower right an alveolar duct can be seen with a number
of adjacent alveoli.   The individual cells which comprise
the walls of the alveolar sacs can be seen along with
several capillaries.   At the lower left the folded
epithelium of a bronchiole can be seen.   A small vessel is
also evident at the top of the image.

     The acoustic image of Fig. 20 shows a section of the
human Fallopian tube.   This area is characterized by the
deep branching folds of the mucous membrane.   The most
striking feature of the image is the sharp contrast between
the internal matrix of connective tissue and the epithelial
layer.   As will be seen in the following figures, the
large acoustic attenuation of connective tissues holds for
a variety of specimen.   In this instance the connective
tissue is loose with numerous fibroblasts.   The columnar
structure of the epithelium is evident in some regions of
the image, and some of the subtle detail within cells can
also be seen.   Of particular interest are the highly

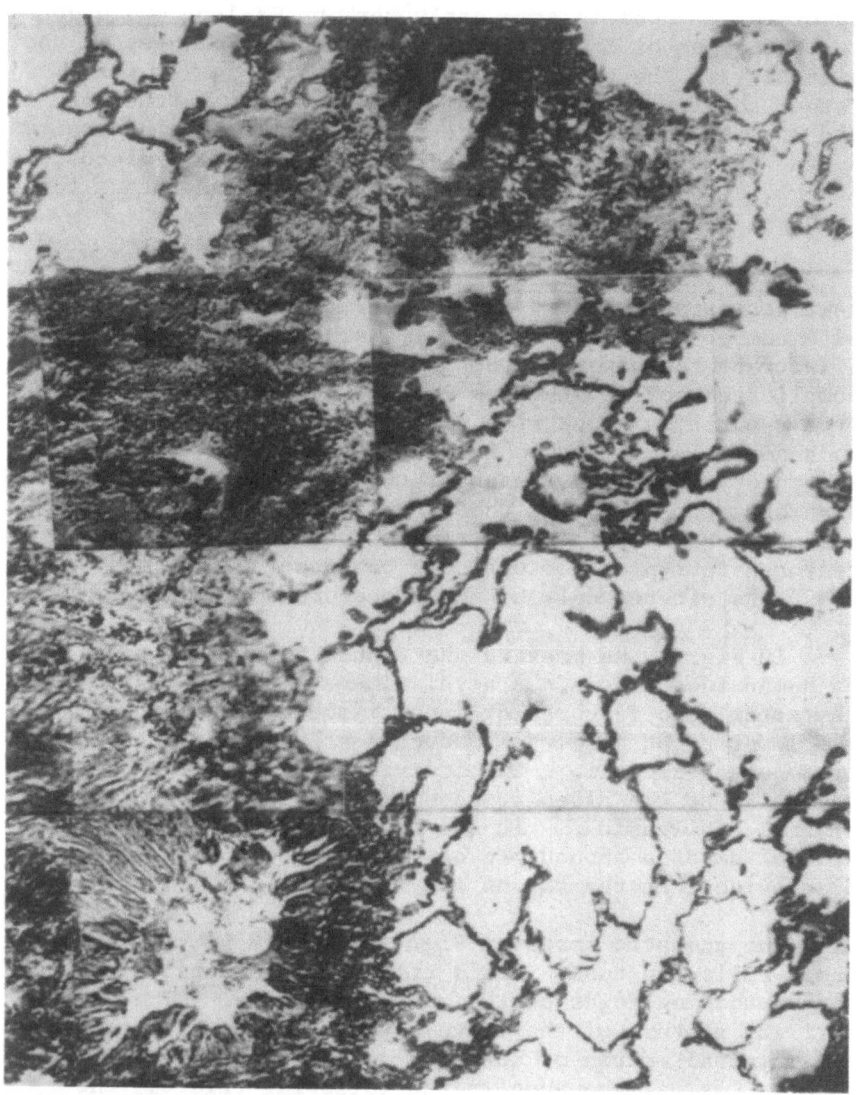

FIG. 19 Acoustic image (600 MHz) of a section of human
       lung tissue (unstained). Magnification ~ 200X.

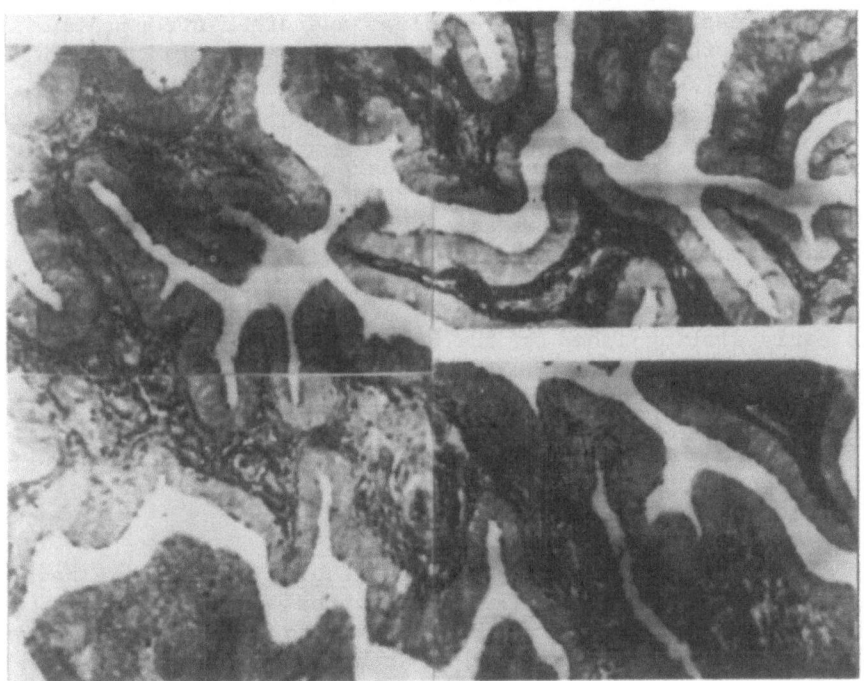

FIG. 20 Acoustic micrograph (600 MHz) of a section of
human oviduct. Magnification ~ 200X.

The surface epithelium is clearly distinguished
from the interior connective tissue which exhibits
strong acoustic absorption and appears dark.
The specimen is unstained.

attenuating regions at the outer boundary of certain cells.
There is evidence to indicate that these cells are ciliated
as opposed to the non-ciliated secretory cells.

Figure 21 is an acoustic micrograph of a section of
human spleen.   The numerous and acoustically absorbing
red blood cells within the tissue are clearly shown as
well as capillaries, fat vacuoles, and other structures.
The most prominent feature is the oblique section of a
small artery.   Inside the vessel a great number of red
blood cells can be seen.   Within the wall of the vessel
the internal elastic membrane is the most distinctive
structure.   This membrane attenuates the acoustic energy
very strongly and thus appears as a folded black line
around the inside of the artery.   In contrast an optical
micrograph shows the elastic membrane as a translucent
structure unless it has been specifically stained.
Surrounding the artery the band of muscle tissue is also
clearly distinguished in the acoustic image.

We are in this way accumulating information on the
acoustic response of various biological structures.   With
it we hope to pinpoint the areas in which the acoustic
microscope will be a valuable tool for research and
diagnosis.

(2) Tissue Sections showing Pathology.   We have been
fortunate in our work on tissue sections for we have
attracted the interest of Dr. R. Dorfman and Dr. R. Kempson
of the Pathology Department at Stanford.   They have taken
time out to identify and furnish us with selected abnormal
tissues of the human breast and of lymph nodes.   We
present some of the results here.

The first acoustic image of this section (Fig. 22)
shows a panoramic view of a malignant tumor of the human
breast.   This carcinoma is of the infiltrating ductile
variety.   After the acoustic images of this specimen had
been recorded the section was stained with hematoxylin and
eosin so that a comparative optical micrograph could be
made.   Thus, Fig. 23 shows an optical image of the same
area shown in Fig. 22.   The micrograph was made with a
Zeiss Photo-Microscope II using a 10X.NA. .32, planapo
objective.

FIG. 21 Acoustic micrograph (600 MHz) of a section of
human spleen showing a small artery (unstained).
Magnification ~ 200X.

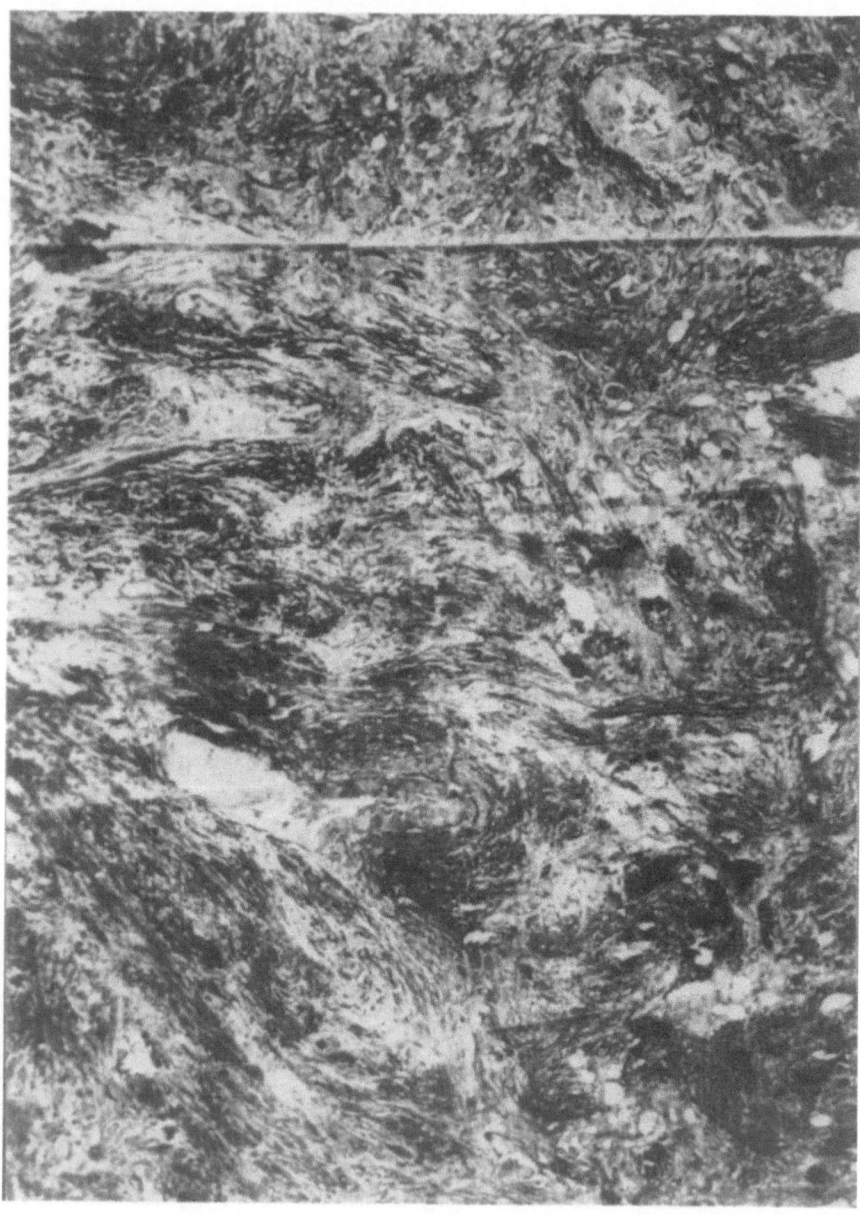

FIG. 22 Acoustic micrograph (600 MHz) of a malignant tumor
of the human breast (unstained).
Magnification ~ 100X.

In order to facilitate a comparison of the detail in the acoustic image with that seen optically an area of interest in Fig. 22 has been selected for enlargement. This enlargement, along with its optical counterpart, is presented in Fig. 24. A close inspection of Fig. 24 will show that the basic structural features seen in the acoustic micrograph correlate on a one-to-one basis with the optical image. For example, at the left of the image a number of neoplastic cells have differentiated into a ductile structure. Each individual cell can be compared in the acoustic and optical images. The emphasis of particular details is, however, quite different. In the optical image the hematoxylin-eosin (HE) stain provides a clear distinction between the cell nucleus and the cytoplasm, while distinctions between cell types are less marked. In contrast differences in acoustic absorption tend to discriminate between cell types. In some instances a difference between two areas can be quite subtle in the optical image while being obvious in the acoustic micrograph.

We are aware that the collagenous regions which show up so clearly as dark areas in Figs. 22 and 24 can also be made clear in the optical micrographs with Masson's trichrome stain. We have made our comparison with HE stained samples for the reason that it takes several hours to prepare the tri-chrome stain. There are important situations, such as those related to biopsies of patients in the operating room, where this interval of time is prohibitive. For those cases where it is necessary to work with frozen sections that are stained with HE we believe that the comparison as shown here is significant. The additional information available from the acoustic micrographs of unstained samples should be of great value - once the research necessary to permit an accurate interpretation of these images has been carried out.

FRESNEL ZONE PLATE AS A LENS

Description of Imaging Methods

The spherical surfaces as used in the microscope perform in superb fashion when used as a lens at high frequencies. There the objects are small in size and the diameter of the lens is compatible with the entire apparatus. However, at the lower frequencies - such as those near 10 MHz

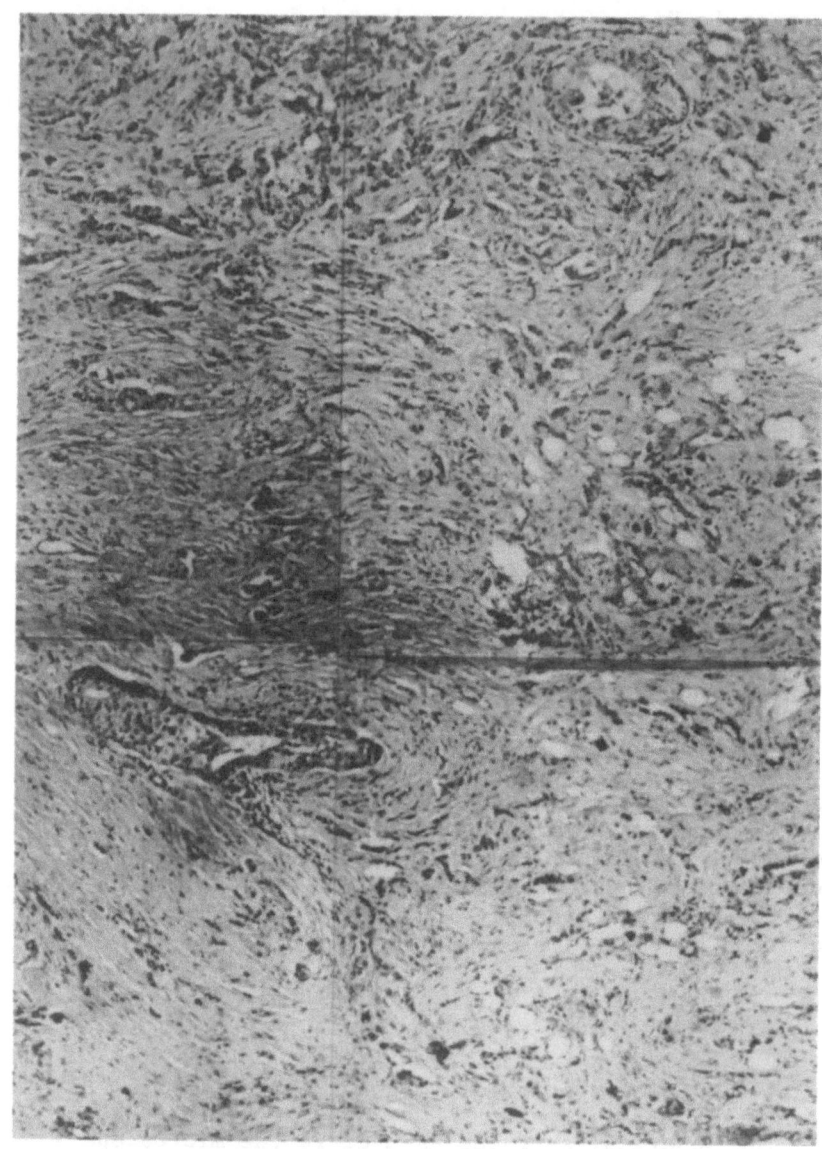

FIG. 23 Optical micrograph of the breast tumor shown in
Fig. 22 (stain: hematoxylin-eosin).
Magnification ~ 100X.

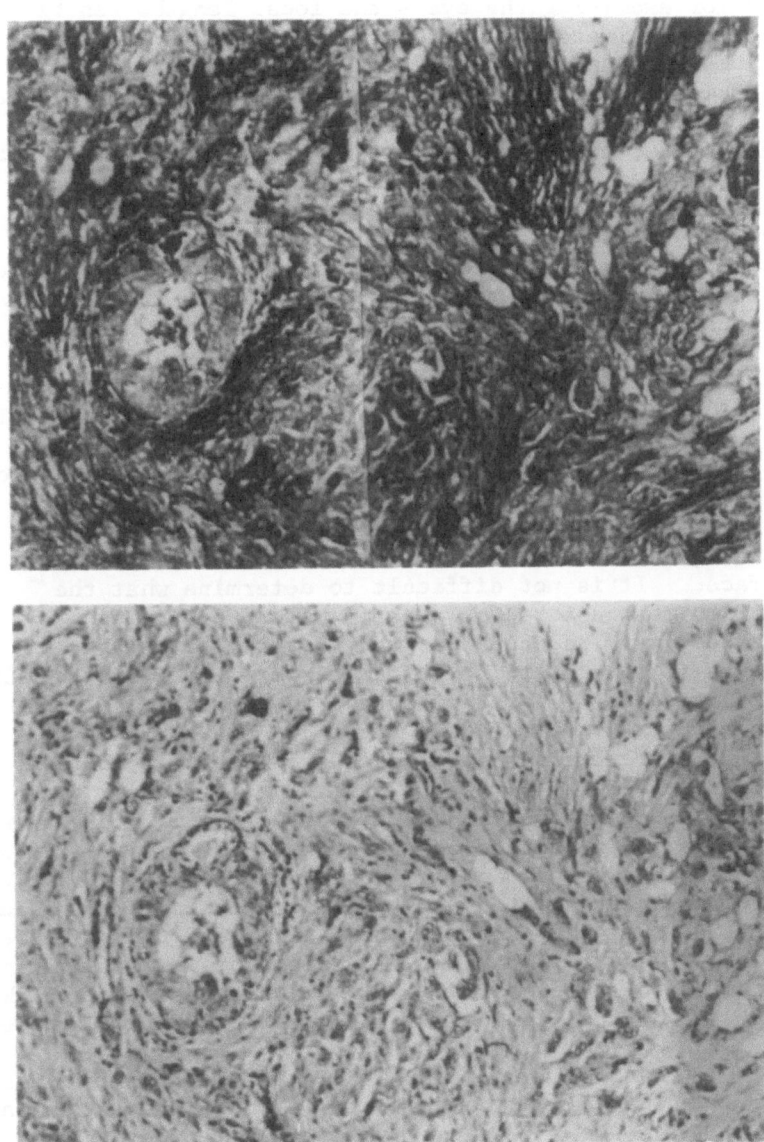

FIG. 24 Comparison of the acoustic (top) and optical
images (bottom) of a small region of the tumor
shown in Figs. 22 and 23.
Magnification ~ 150X.

used for examining the eye - the focal lengths can be
several centimeters and the lens size for F/1 systems
can be large and inconvenient.   In this situation a
Fresnel zone plate, or more properly a Fresnel phase plate,
does have advantages.   It comes in the form of a thin
plate rather than a sphere and it has a focal length that
can be varied electronically by changing the frequency.
These two features have attracted our interest as a
possible technique for realizing an acoustic imaging
system that is useful for examining selective regions such
as the human eye.   We want to describe the work that has
been carried out up to the date of this writing (Spring
of 1975) by S. Farnow working with B. A. Auld.[52]

   There are a variety of ways to present the principles
of a Fresnel lens but the one that appeals to us can be
described as follows.   We use a plane surface in the zone
plate to generate a spherically converging wave.   This can
be done if we control the phase and amplitude of the
exciting source over the entire surface of the plane
surface.   It is not difficult to determine what the
distribution must be.   We begin with a diverging
spherical wave and consider the phase and amplitude of this
wave as it intersects the plane surface - the surface that
will act as our generating source in the lens.   The
situation is illustrated in Fig. 25.   We can see from
that illustration that the regions of constant phase are
circles of varying diameter.   The wave amplitude does
decrease somewhat as the radius increases but the change
is not large in most cases and it will be neglected.   With
this in mind we have only to arrange matters such that we
excite the plates with rings of alternating phase.   The
actual size of the rings can be determined without undue
labor by requiring that the wavelets from alternate rings
reach the focal point in phase.   [For those with a background in optics, this relates to the Huygen's principle.]

   The mathematical relations that are used in the design
of the zone plate follow from this requirement of constant
phase.   See Fig. 26 and note that the path length from
the successive zones (i.e. rings) to the focal point is
given by the relation

$$r_n^2 = R_n^2 - F^2 \tag{5}$$

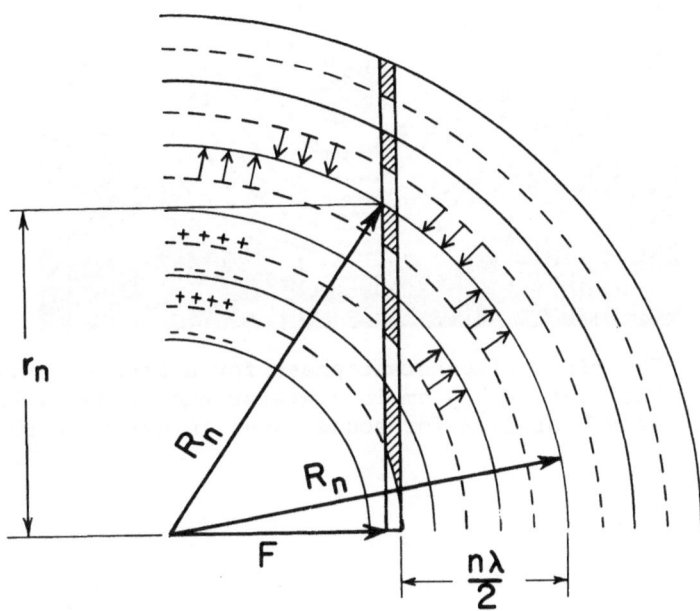

FIG. 25 The circles of constant phase for a spherically diverging wave front. Here the solid circles represent the crests of the pressure wave and the dotted circles represent the troughs. Thus these successive circles are spaced by one-half wavelength. The cross hatched regions on the vertical plate are the regions which should be excited in phase if this plane surface is to excite a spherical wave. The alternate region between the cross hatching should also be excited in phase but with a polarity that is opposite to the cross hatched regions.

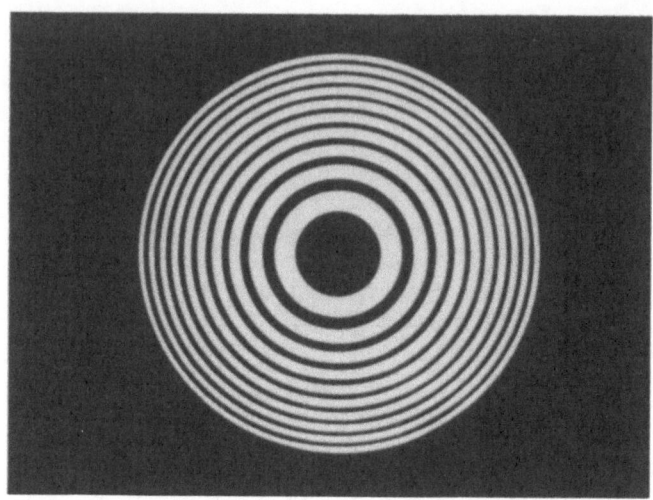

FIG. 26 The rings of constant phase for a Fresnel lens.
The lens is 1.9 cm in diameter and it has a focal
length of 3 cm for sound waves in water at 10 MHz.

but we, also, note that

$$R_n = F + n\lambda/2 \tag{6}$$

therefore, eq. (5) takes the form

$$r_n^2 = F^2 + n\lambda F + n\lambda \left(\frac{n\lambda}{4}\right) - F^2$$

or

$$r_n^2 = n\lambda \left(F + \frac{n\lambda}{4}\right) \tag{7}$$

In the typical cases that will be of concern to us
the value of the focal length F is 200λ and with a ten
zone plate ($n = 20$) we see that $\frac{n\lambda}{4} \ll F$. We can,
therefore, write eq. (7) in the form

$$r_n^2 \approx n\lambda F \tag{8}$$

This is the relation that is used to design the rings on
the piezoelectric plate.   The actual ring pattern as seen
from the front face is shown in Fig. 26.   In this pattern
we excite the dark rings with a phase that is opposite to
that of the white rings.   The phase of the acoustic wave
is determined by the product of the r.f. electric field
across the plate and the poling direction within the plate.
We, therefore, have a choice - we can either change the
phase of the electric field from ring to ring - or we can
excite the plate with a uniform electric field and fabricate
the plate with alternate poling in the adjacent rings.

As it turns out the latter method has a number of
advantages.   The r.f. field configuration where it is
uniform throughout the plate is easily applied and it is a
more efficient use of the system in that a greater fraction
of the electromagnetic energy is converted to acoustic
energy.   Furthermore, it is a relatively easy matter to
achieve the alternate poling rings.   This is done by
depositing a series of rings in the form of metal films on
the ceramic plate with the geometry shown in Fig. 26.
A d.c. voltage is applied across the plate and the entire
assembly is carefully heated to $150^{\circ}C$.   At this
temperature the poling direction in those regions with an
applied d.c. field reverses in such a way that the polarization aligns itself with the d.c. field.   As it cools to
room temperature the polarization is fixed, the rings are
removed and replaced with a uniform metal electrode.   The
polarization within the plate has the characteristic shown
in Fig. 27.

An r.f. electric field across the uniform electrodes
will now excite an acoustic field in the form of a
converging spherical wave.   This wave will converge to a
focus as given from eq. (8)

$$F = r_n^2/n\lambda = \frac{r_n^2}{nv_s}\text{ freq.}\qquad(9)$$

where v_s is the velocity of sound.   Thus we learn that
the focal length F in such a configuration can be varied
directly as the r.f. frequency.   This is an important
feature of this configuration for it allows us to work with
an "electronic scan" along the dimension of the axis.   A
plot of the focal length versus frequency is shown in

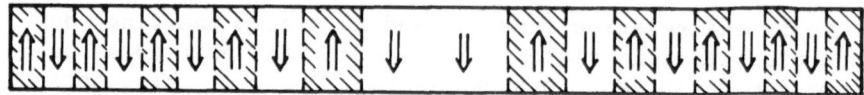

FIG. 27 The alternate poling direction in the piezoelectric
        ceramic. With this configuration an r.f. electric
        field applied uniformly over this plate will excite
        the regions with the upward arrow opposite in
        phase to those with the downward arrow. Thus we
        meet the conditions indicated in Fig. 25 for
        generating a spherical wave.

Fig. 28 in order that the reader may appreciate the
scanning distances that are achieved when the frequency is
varied.

     The resolution of this type of imaging system is
determined by the diameter of the beam at the focal point
and in the lens as used by Farnow the beam contour at the
focal point is given in Fig. 29. We see that the beam
width is 0.3 mm which is approximately 2λ at the
frequency of 10 MHz where the focal length is 3 cm.

     As a final item before we go into a discussion of the
images we point out that this type of lens when used as a
receiving system can respond only to spherical waves
diverging from a point at a distance F from the lens
surface.   It cannot detect plane waves.   Therefore, when
used in an imaging system the phase plate - responsive to
radiation that is scattered from the object in the focal
plane - discriminates against the waves that originate
from other points.   We believe that this is an advantage
in reducing the artifacts that can arise with other
imaging systems with objects that are not in the focal
plane.

                    Imaging System and Results

     The heart of any imaging system is the lens and with
the Fresnel phase plate as described above the overall
imaging system takes on many of the characteristics

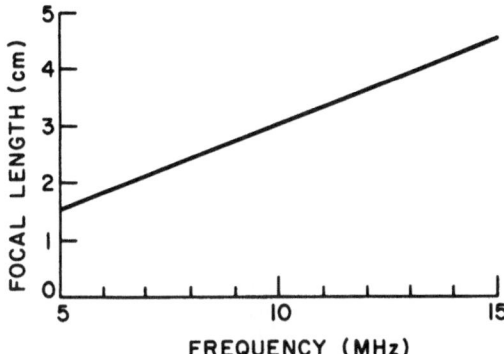

FIG. 28 The relationship between focal length and
frequency. By varying the frequency from
5 to 15 MHz, the focus may be scanned 3 cm
along the axis of the transducer.

described in connection with the acoustic microscope.

Much of the work has been done with a fixed frequency
and with slow scanning in two dimensions with mechanical
methods. Both the lens system and object are immersed in
a water bath. The scanning is achieved by scanning either
the object or the lens. In the transmission mode two
confocal lens are placed on either side of the object.
One is used to illuminate this object and the other
receives the energy that is scattered after transmission
through the specimen. A typical image obtained with this
technique is shown in Fig. 30.

The reflection mode is also conveniently adapted to
this system. Here a single lens is used both as an
illuminator and a receiver for the back scattered energy.
Typical results for this type of imaging are presented in
Fig. 31 and Fig. 32.

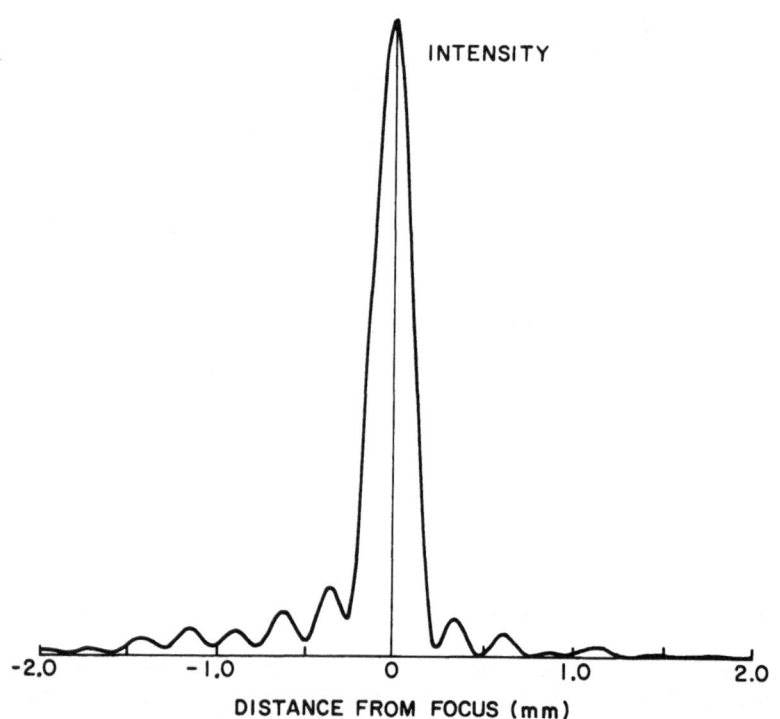

FIG. 29 Focal plane intensity distribution for the APP
transducer, operating at 10 MHz with a 3 cm focal
length.   The left sidelobes are slightly enhanced
due to a slight overlap of the disc electrode on
that side of the APP transducer. (Farnow).

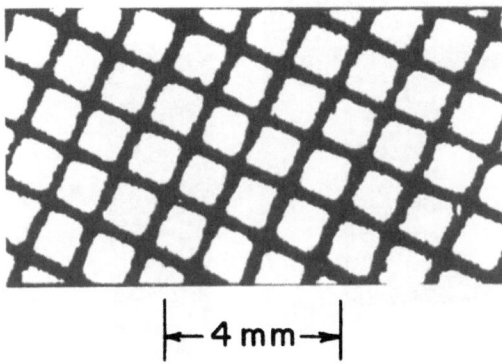

FIG. 30 Acoustic transmission image of wire mesh made
with the APP transmitter.   (Farnow).

In each of these illustrations the lenses were held
fixed and the object was mechanically scanned.   We have
obtained similar images holding the object fixed and
moving the lenses.   In principle this is not different
from the previous situation but in practice it is
important to maintain the object fixed - very often it is
attached to a larger body and it is not convenient to
move it in the regular pattern required for scanning.
Further than this, we can with a fixed object change the
focal length with frequency shifts and obtain the image
with mechanical scanning in one-dimension.   The electronic
scanning along the axis of the lenses can be very fast and
the mechanical scanning of one-dimension can be carried
out at a rate of 60 Mz.   Thus, the images can be
displayed on a TV monitor in real time.   We are of the
opinion that this will one day prove to be of great
advantage in a new instrument that serves as an "acoustic
ophthalmoscope".

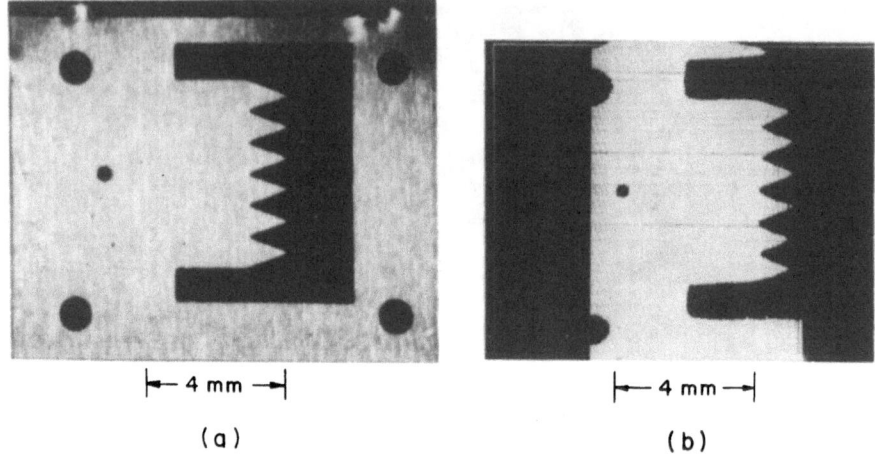

FIG. 31 Comparison of (a) optical and (b) acoustic
reflection images of a sawtooth pattern punched
into a 3-mil nickel-plated copper sheet. (Farnow).

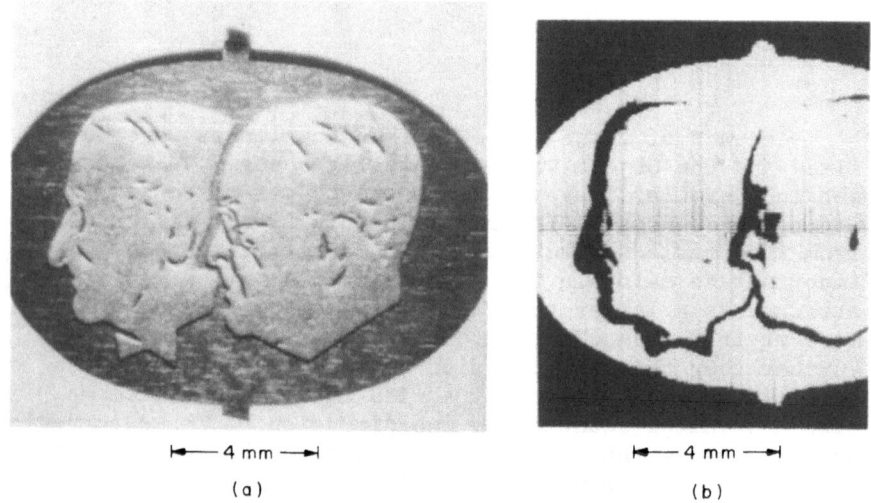

FIG. 32 Comparison of (a) optical and (b) acoustic
reflection images of a relief pattern etched
on a 3-mil nickel-plated copper sheet. (Farnow).

CYLINDRICAL ELECTRONIC LENS

The concepts that were used in connection with the Fresnel lenses of the previous section can be extended to provide us with an imaging system with still greater versatility if we move to cylindrical geometry. There the fronts of constant phase are cylinders and the intersections of these wavefronts with a plane form straight lines (Fig. 33). The lens in our new system will consist of a linear array of piezoelectric elements mounted on this plane. We will show that it is possible to sum the outputs from each element of this linear array in such a way that both scanning and focusing can be carried out with electronic rather than mechanical techniques. We will use the term "electronic lens" for this system.

We will proceed step-by-step in this description so that the principle that we wish to employ can be brought out with a minimum of algebra as mentioned above. We will work with a linear array of piezoelectric elements (the reasoning behind this choice becoming clear when we come to the section on scanning) and we want to use this linear array to efficiently detect the acoustic energy that comes from a line source in the form of a cylindrically diverging wave. The obvious way of doing this is to use a lens made of high velocity material - a lens with a focus centered on the line source. This will convert the diverging wave into a plane wave and each element of the array will be excited in phase. We then merely sum the signal from each to maximize the output. We can see how this fits together from the sketch in Fig. 34.

But this function of compensating for the unequal path length can be realized in quite a different way. If we accept the output of each array element as it is obtained from the diverging wave we will find that the phase from adjoining elements will vary. This variation is such that if the elements were summed directly the output would be very small as compared to that from the arrangement of Fig. 34. We, therefore, choose not to sum the outputs directly but rather we will use an alternate mode. We first mix the signals with a second frequency and sum the outputs at the difference frequency. In this situation the phase at each point on the array for the difference frequency is equal to the differing in the phase θ from the array and the phase φ from the second frequency.

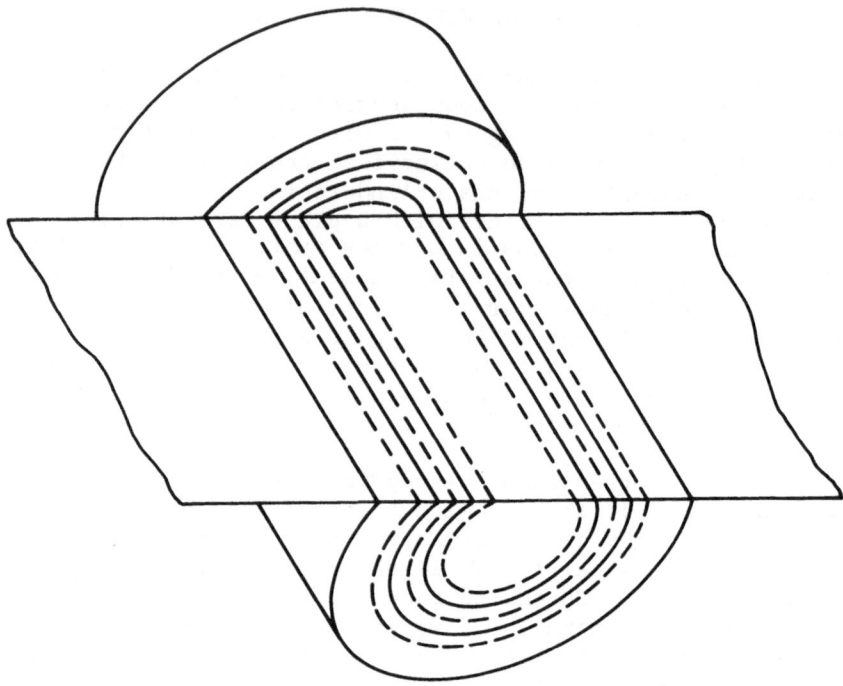

FIG. 33 Illustration of the intersection of a cylindrically diverging wave on a plane surface.
The surfaces of constant phase intersect the
plane as straight lines.

We then adjust the phase φ at each mixing point in such
a way that it exactly compensates for the variation in θ ,
the phase of the incoming signal wave.   Thus the term
$\varphi\theta$ is constant.   When this is summed over the array
elements we again reach the maximum in the output as
before.

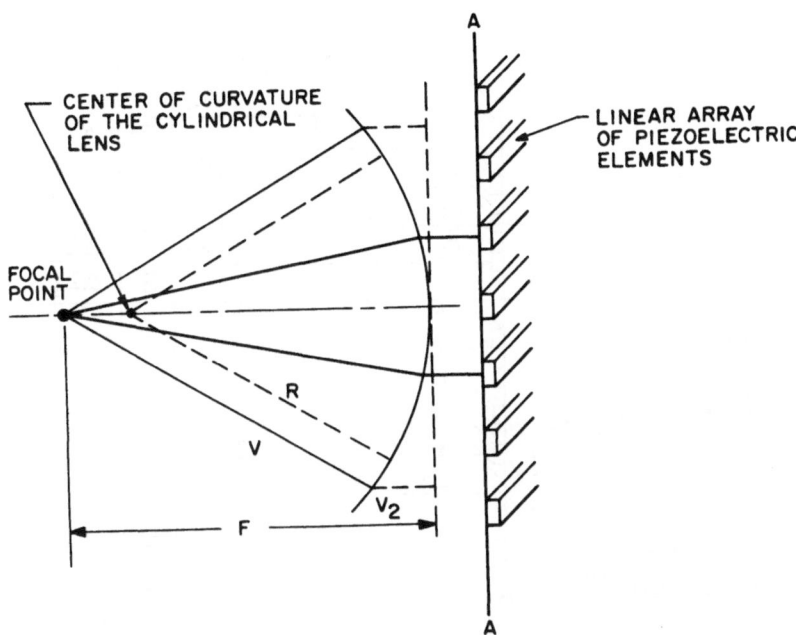

FIG. 34 A cylindrical lens of high velocity material uses
for the purpose of compensating for the additional
path length of those rays which leave the object
at a wide angle.   With this lens the wave arrives
at the surface A-A in phase and the elements of
the array can be directly summed to give a maximum
output for a line source located as shown.

       With the aid of the sketch in Fig. 35 we can illustrate
a rather simple way in which the desired variation in θ
can be realized.   The second frequency is obtained from
taps on an acoustic surface wave delay line.[33]   Although
this is still not the version that is to be incorporated
in the final system it does facilitate the explanation of
that system.   We can calculate the variation, $\Delta\theta$, in
the phase of the incoming signal wave for two adjacent
array elements located at distance x from the central
axis.   Here Δx is the spacing of the elements and the

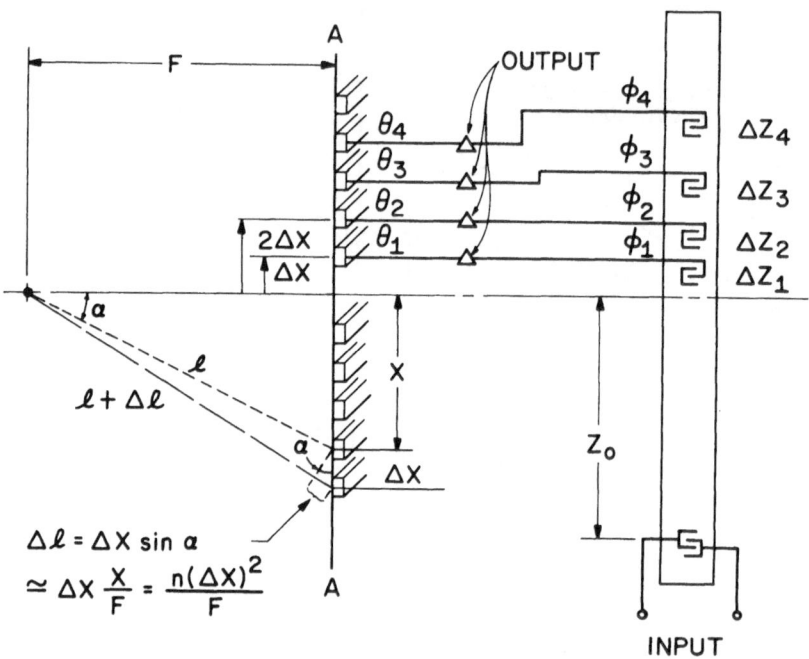

FIG. 35 A system for focusing that achieves the same
        result as the lens of Fig. 33. Here the output
        is at $\omega_0 - \omega_s$ and the phase at each tap is $\theta - \varphi$.
        By proper spacing of the taps on the delay line
        the variation in φ can match the phase variation
        in θ and the term $\theta - \varphi$ remains constant.

difference in path length, Δl , is given by

$$\Delta l = \Delta x \, \sin \alpha = \Delta x \, \frac{x}{R} \tag{10}$$

but for $x \ll R$ we have $R \sim F$

    There is, of course, another way of compensating for
the variation in φ . We could insert a delay line in
each lead coming from the array elements and individually
adjust the delays to compensate for the variations.

This is a rather direct electronic analog to compensation as accomplished by the lens of Fig. 34.   Such a system has been described earlier in Chapter 7 by Macovski. The alternative described here as based on the concept of frequency shifting will in principle perform that same function as those systems which use individually controlled phase shifts at each element.   The advantages and disadvantages involve a point-by-point comparison of the details of the electronics that must be incorporated to implement the schemes.   This comparison has not yet been made and it will probably have to wait until the various systems have been more widely used before the comparison can be done with accuracy.

The difference in phase between two adjacent elements is given by

$$\Delta\theta = \frac{\omega_s}{v_s} \; \Delta l = \frac{\omega_s}{v_s} \; \Delta x \, \frac{x}{F} \tag{11}$$

For our acoustic surface wave delay line the difference in phase, between two adjacent taps is simply

$$\Delta\varphi = \frac{\omega_m}{v_m} \; \Delta z \tag{12}$$

To maximize our output signal we adjust the location of the taps so that $\Delta\varphi = -\Delta\theta$ or

$$\frac{\omega_m}{v_m} \; \Delta z \; = \; \frac{\omega_s}{v_s} \; \Delta x \; \frac{n\Delta x}{F} \tag{13}$$

Here we are interested in the nth element where $x = n\Delta x$ and then

$$\Delta z = \frac{\omega_s}{\omega_m} \; \frac{v_m}{v_s} \; \frac{(\Delta x)^2}{F} \; n \tag{14}$$

Thus, we see that the spacing Δz increases from tap to tap directly as n since all other factors on the right of eq. (14) are constant.

With this adjustment of $\Delta\varphi$ we are left with the problem of arranging the output circuitry so that it is responsive to the difference of $(\Delta\theta - \Delta\varphi)$.   To do this

we mix the output signal from the nth tap of the array
with the mixing frequency from the nth tap on the delay
line.   The output from these n mixing elements at the
difference frequency contains the phase information
$\Delta\theta - \Delta\varphi$.   Since this is constant we can sum the outputs
directly to obtain a maximum similar to that of the
arrangement in Fig. 34.

It is well to ask at this point - "why bother with
this alternative ?"   It is a good question for the system
is responsive only to emission of acoustic energy from a
single line.   It does not permit scanning and it has no
advantage of the lens of Fig. 34, but it does serve to
extend our thinking.   With one modification we can
describe a system that will permit us to scan along a line
parallel to the receiving array and control the focal
length - both of these by electronic means.   To understand
the new modification we can return to eq. (12) $(\Delta\varphi = \omega_m \Delta z / v_m)$.
Our requirement for the "electronic lens" is that we adjust
the value of $\Delta\varphi$ at the nth tap to correspond to the
variation of $\Delta\theta$ expressed by eq. (12).   As a first step
we chose to vary Δz - which meant varying the spacing of
the taps - but this can only be done in a fixed way by
fabricating a special pattern on the delay line.   We can
equally well vary the other two parameters on the right of
eq. (12) - that is either v_m or ω_m .   A variation of
v_m is difficult but a variation of ω_m is straightforward.
All we require is a proper variation of ω_m with time at
the input to the delay line.   It is easy to specify this
variation.[34]

If we inject a frequency ω into the delay line at a
time t_o it will arrive at the nth tap (a distance
$z_o + n\Delta z$ from the input) at a time given by

$$t_1 = t_o + \frac{z_o + n\Delta z}{v_\ell} \qquad (15)$$

Here t_o is the time of injection at the input and t_1
is the arrival time at the nth tap.   We propose to
vary the frequency at the input in accordance with the
relation $\omega = \omega_o + \mu t_o$.   We will choose μ to be
constant and, therefore, ω increases linearly with time.
At the nth tap we can find the frequency by replacing
t_o by $t_1 - (z_o + n\Delta z)/v_\ell$ to obtain

$$\omega_m = \omega_o + \mu_o t_1 - \mu z_o/v_\ell - \mu n\Delta z/v_\ell \qquad (16)$$

If we choose a particular time

$$t_1 = z_o/v_\ell \qquad (17)$$

this reduces to the simple expression

$$\omega_m = \omega_o - \mu n\Delta z/v_\ell \qquad (18)$$

We recall that the phase shift between adjacent taps is given by $\Delta\varphi = \omega_m \Delta z/v_\ell$ and therefore

$$\Delta\varphi = \omega_o \Delta z/v_\ell - \mu n(\Delta z/v_\ell)^2 \qquad (19)$$

With this variation we have achieved our goal and the arrangement is sketched in Fig. 36. The difference between the phase shift $\Delta\theta$ from the array and $\Delta\varphi$ from the taps can be constant. We subtract eq. (19) from eq. (11) to find

$$\Delta\theta - \Delta\varphi = \omega_o \Delta z/v_\ell \qquad (20)$$

This is constant since the tap spacing, Δz is fixed. Eq. (20) is true if we arrange matters such that

$$\mu \ (\Delta z/v_\ell)^2 \ n = (\omega_s v_s/F) \ (\Delta x/v_s)^2 \ n$$

or

$$\mu = (\omega_s v_s/F) \ (\Delta x v_\ell/\Delta z v_s)^2 \qquad (21)$$

The parameter μ which denotes the rate of increase in frequency is known as the "chirp" rate. The number of taps on the delay line is equal to the number of elements in the array and we see that $\Delta x/\Delta z$ is simply the ratio of the breadth of the array to the length of the delay line. This typically is ten. In a usual case $v_s \sim v_\ell$ and $(\Delta x v_\ell/\Delta z v_s)^2 \sim 100$. For a system operating in water $(v_s = 1.5 \times 10^5$ cm/sec) with a focal length F of 20 cm a signal frequency of 1.5 MHz and a mixing frequency of 100 MHz we can calculate the value of μ to be 10,000 ω_m. Thus, in one second of time the frequency shift would be

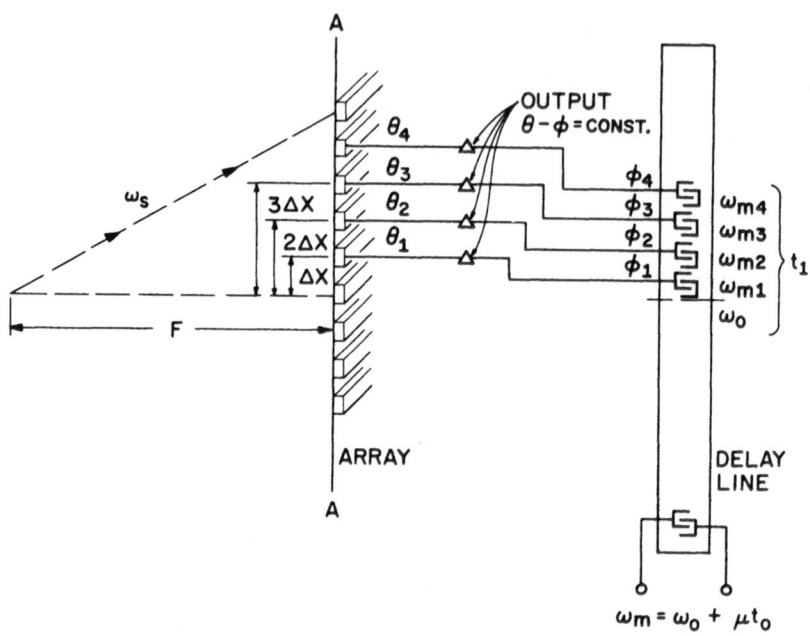

FIG. 36 A variation of the focusing system sketched in
Fig. 34. Here the tap spacing is fixed at Δz but
the frequency is changed in a linear fashion such
that the phase shift $\varphi = \omega_m z/_v$ can be varied in
a way that will keep the term $\theta - \varphi$ constant from
tap to tap.

$10^4 \omega_m$ and in one r.f. cycle $(10^{-8}$ sec) the frequency
shift would be $10^{-4} \omega_m$. This is a perfectly reasonable
"chirp". We see that the entire system fits together if
we use parameters near these values.

We can use these relations to point out the two major
features that make this system so interesting. First, we
see from eq. (17) the system is a scanning system. It
responds to energy emitted by a line source and arriving

at the array at the time t_o $(= z_o/v_\ell)$. Since the taps on
the delay line have a one-to-one correspondence to the
elements of the array this corresponds to acoustic energy
emitted from a line source located at x_o in the lateral
dimensions and placed at distance F from the array.
But as time increases the value of x_o increases as $v_\ell t_o$.

In another form we can state that the system with the
chirped waveform centered on the nth element can receive
energy from a point source located opposite to the nth
element of the array. The chirped waveform is moving
along the delay with the velocity of the acoustic surface
waves. It, therefore, allows the system to scan a series
of sources in the focal plane at this rate. It is an
electronic system (without switches) for one-dimensional
scanning as illustrated in Fig. 37.

The second dimension can be obtained in one of two
ways. We can use mechanical scanning of the transducer
array in the other dimension to obtain a "C-scan", of the
image in the focal plane parallel to the plane of the array.
Or, we can change the focal length, F, from line-to-line
to obtain a "B-scan" of the image in a plane normal to the
plane of the transducer array. The focal length is
adjusted electronically by changing the value of μ, the
chirp rate. We can see this from eq. (21) for there we
find that the focal length is inversely proportional to the
chirp rate, μ. This forms a two-dimensional electronic
scanning system that permits viewing in "real-time" - that
is television frames rate where the flicker is reduced to
a value that is not objectionable.

The object must be properly illuminated with sound and
this can be accomplished with a similar system. It is only
necessary to interchange the input and output from the array.
Thus, we apply an input frequency to the series of mixers
and this when beat against the moving chirp on the delay
line feeds each piezoelectric element in the acoustic
array with a phase proper for focusing at a distance F.
This transmitting array and the receiving array can be
located on opposite sides of the object to form a transmission system or on the same side of the object to form
a reflection system. The system that is employed will
depend on the nature of the object.

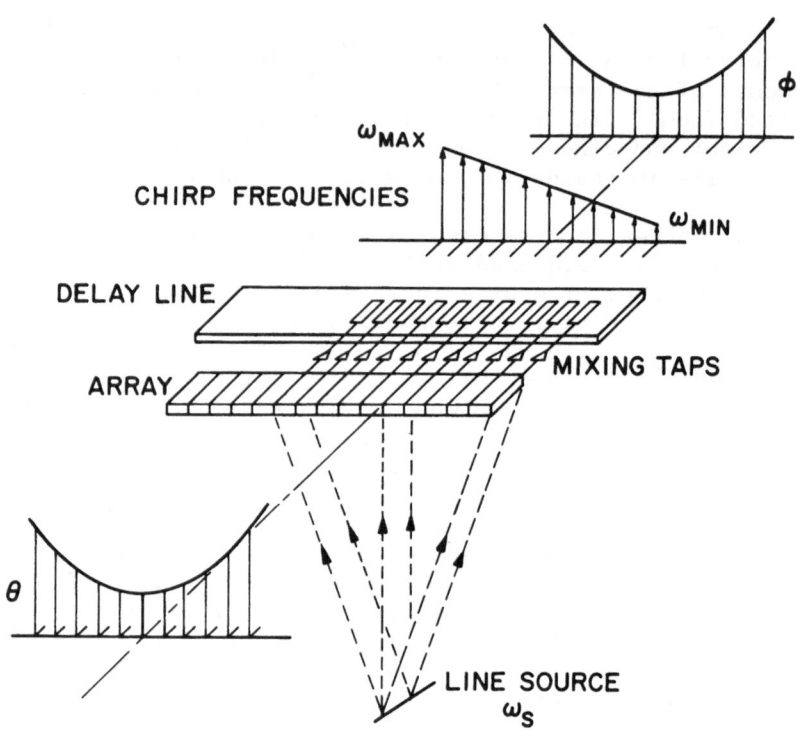

FIG. 37 A composite representation of the electronic
focusing system. The phase variation of the
spherical wave impinges on the array is represented
by the θ bars. A similar variation in phase at the
taps on the delay is represented by the φ bars.
This results from the chirp frequencies extending
from ω_{min} to ω_{max}. The difference in phase between
θ and φ as recorded at the mixing taps is constant.

Some Results

Some of the initial images obtained with this system
have appeared in the literature.[34] The systems that have
been tested are not as fully developed as those described
in the earlier chapters but the results are exciting to
those that are engaged in the work for they are convinced
of the virtues of this system for acoustic imaging. It
does appear that the circuitry using this form of analog

signals is less complex than a system which converts the signals to digital form before processing.[35]

We can illustrate the possibilities with a "C-scan" picture in Fig. 38 and a "B-scan" picture in Fig. 39.

The "C-scan" picture was obtained by electronically scanning the horizontal dimension along the array and mechanically scanning the transducer in the vertical dimension. The field of view is approximately 8 cm by 24 cm. The object is a test piece for a non-destructive test program. It consists of a laminated piece of material which contains defects in the bonding internal to the specimen. These defects are clearly delineated in the photo in Fig. 38.[36]

In the second photo of Fig. 39[37] we have a "B-scan" display of a series of milled steps on an aluminum block. The scan was "all electronic". The super-imposed optical photo is included to give the reader a clear idea of just what it is that we are seeing in the acoustic image.

The "B-scan" photo was recorded with a rate of 30 frames a second whereas the "C-scan" photo with mechanical scanning required 6 seconds for a single frame. The field of view is much larger in this second case and the frame rate is adequate for a large number of applications.

Final Remarks on Lens Systems

The selection of acoustic imaging systems that were included in this chapter was somewhat arbitrary. There are other systems[38] which employ "electronic means" for focusing and scanning and there will be other systems proposed in the near future. Nonetheless, the three systems described here, when coupled with the systems detailed in the earlier chapters, can give the reader some conception of the variety of ideas that are being pursued to determine the optimum system that will eventually be used for imaging with acoustic waves.

The frequency range in these systems is large, extending from 1 MHz in those systems designed to explore the body cavity to 1000 MHz in the microscope as used to examine the interior of individual cells. But these are

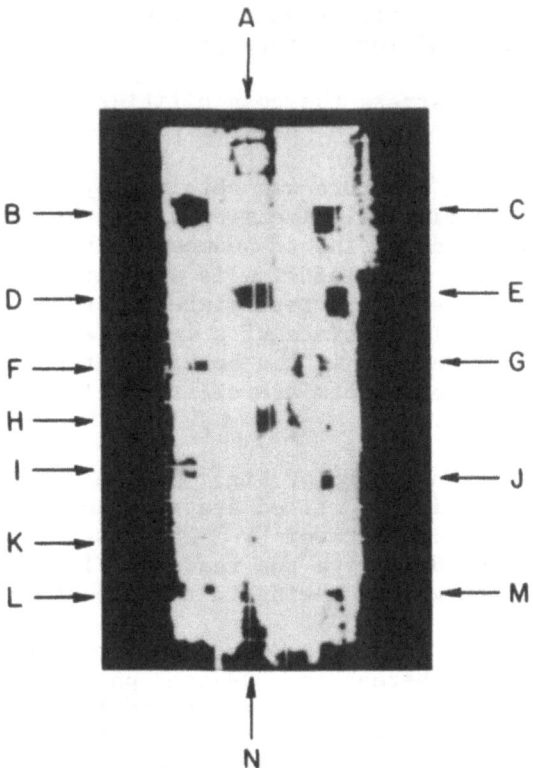

FIG. 38   The acoustic image of a composite boron fiber
          epoxy laminate bonded to a .09" titanium plate.
          The laminate 4 mils in thickness was made of
          15 layers.   The regions of improper bonding
          A, B, C, - etc. are quite evident.

still not limiting for in the future the electronic lens
concept will surely be extended to lower frequencies in a
system designed to explore huge structures such as those
found underground.   And the range of frequencies used in
the microscope will be gradually increased until comparable
with the work now being done with high quality optical
microscopes.

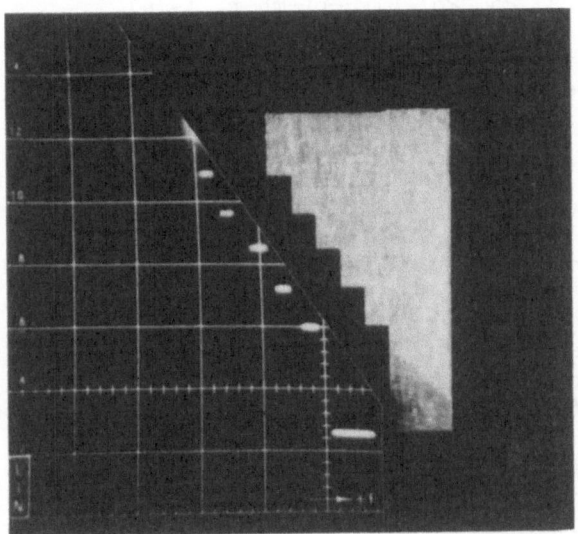

FIG. 39 The acoustic photo of the steps in the metal
block shown as an optical photo on the right.
This is the reflection mode for the electronic
lens and the reflecting points at each step
are clear in the "B-scan" form of presentation.

REFERENCES

1. Lemons, R. A. and Quate, C. F. "Acoustic Microscopy:
   Biomedical Applications", Science, 188, 905
   (May 30, 1975). Copyright 1975 by the American
   Association for the Advancement of Science.

2. Lemons, R. A. and Quate, C. F. "Integrated Circuits as
   Viewed with an Acoustic Microscope", Appl. Phys.
   Letters, 25, 251 (1 September, 1974).

3. Dunn, F. and Fry, W. J. "Ultrasonic Absorption
   Microscope", J. Acoust. Soc. Amer., 31, 632 (1959).

4. Encyclopedia of Microscopy, Chapter entitled "Ultrasonic Microscope", G. L. Clark, Editor, Reinhold,
   London (1961).

5. Sokolov, S. "The Ultrasonic Microscope", Akademia Nauk
   SSSR, Doklady, 64, 333 (1949).

6. Carstensen, E. L., Li, K. and Schwan, H. P. "Determina-
   tion of the Acoustic Properties of Blood and its Components", J. Acoust. Soc. Amer., 25, 286 (1953).

7. Carstensen, E. L. and Schwan, H. P. "Acoustic Properties of Hemoglobin Solution", J. Acoust. Soc. Amer.,
   31, 305 (March 1959).

8. Kessler, L. W. "VHF Ultrasonic Attenuation in Mammalian
   Tissue", J. Acoust. Soc. Amer., 53, 1959 (1973). See
   also, Goldman, D. E. and Hueter, T. F. "Tabular Data of
   the Velocity and Absorption of High Frequency Sound in
   Mammalian Tissue", J. Acoust. Soc. Amer., 28, 35 (1956).

9. Lees, S. and Barber, F. E. "Looking into the Tooth and
   its Surfaces with Ultrasonics", Ultrasonics, 9, 95
   (April 1971).

10. Anderson, R. E. "Potential Medical Applications for
    Ultrasonic Holography", in Acoustical Holography, 5,
    505, P. S. Green, Editor, Plenum Press, New York (1974).

11. Korpel, A., Kessler, L. W. and Palermo, P. R. "Acoustic
    Microscope Operating at 100 MHz", Nature, 232, 110
    (1971).

12. Auld, B. A. <u>et al</u>. "A 1.1 GHz Scanned Acoustic Microscope", in <u>Acoustical Holography</u>, 4, 73, G. Wade, Editor, Plenum Press, New York (1972).

13. Thompson, J. K., Wickramasinghe, H. K. and Ash, E. A. "A Fabry-Perot Acoustic Surface Vibration Detector - Application to Acoustic Holography", J. Phys. D, Appl. Phys. 6, 677 (1973).

14. Cunningham, J. A. and Quate, C. F. "Acoustic Interference in Solids and Holographic Imaging", in <u>Acoustical Holography</u>, 4, 667, G. Wade, Editor, Plenum Press, New York (1972). See also, Cunningham, J. A. and Quate, C. F. "High-Resolution High-Contrast Acoustic Imaging", J. Physique, 33, Colloque C-6, Supplement, 42 (Nov-Dec. 1972).

15. Mueller, R. K. "Acoustic Holography", Proc. IEEE, 59, 1319 (1971).

16. Kessler, L. W. "Review of Progress and Applications to Acoustic Microscopy", J. Acoust. Soc. Amer., 55, 909 (1974).

17. Korpel, A. "Acoustic Microscopy", in <u>Ultrasonic Imaging and Holography</u>, G. W. Stroke <u>et al</u>. (editors), 345-362, Plenum Press, New York (1974).

18. Cunningham, J. A. and Quate, C. F. "High-Resolution Acoustic Imaging by Contact Printing", in <u>Acoustical Holography</u>, 5, 83, P. S. Green, Editor, Plenum Press, New York (1974).

19. Kessler, L. W., Palermo, P. R. and Korpel, A. "Recent Developments with a Scanning Laser Acoustic Microscope", in <u>Acoustical Holography</u>, 5, 15, P. S. Green, Editor, Plenum Press, New York (1974).

20. Lemons, R. A. and Quate, C. F. "A Scanning Acoustic Microscope", 1973 Ultrasonics Symposium Proceedings, IEEE Cat. #73CHO807-8 SU, 18-21 (1974).

21. Lemons, R. A. and Quate, C. F. "Acoustic Microscope - Scanning Version", Appl. Phys. Letters, 24, 163 (15 February 1974).

22. Rand, S. C. and Stoicheff, B. P. "Brillouin Line Width
     and the Attenuation of Sound in Liquid Argon", Physics
     Letters, 48A, 355 (15 July 1974).

23. Herzfeld, K. F. and Litovitz, T. A. in Absorption and
     Dispersion of Ultrasonic Waves, Academic Press, New
     York (1959).

24. Squires, C. F. in Waves in Physical Systems, Sect. 6.2,
     Prentice Hall, New Jersey (1971).

25. Lamb, J. "Thermal Relaxation in Liquids", in Physical
     Acoustics, II-A, 203, W. P. Mason, Editor, Academic
     Press, New York (1965).

26. Gwertz, S. et al. "Brillouin Spectra of Ethyl Ether
     and Carbon Disulphide", J. Acoust. Soc. Amer., 49,
     994 (1971).

27. Attal, J. Private communication.

28. Hall, L. "The Origin of Ultrasonic Absorption in
     Water", Physical Review, 73, 775 (1 April, 1948).

29. Stuehr, J. and Yeager, E. "The Propagation of
     Ultrasonic Waves in Electrolytic Solutions", in
     Physical Acoustics, II-B, 351, W. P. Mason, Editor,
     Academic Press, New York (1965).

30. Breitschwerdt, K. G. and Kistenmacher, H. "Ultrasonic
     Absorption and Molecular Motions in Ionic Solutions",
     Journal of Chemical Physics, 56, 4800 (1972).

31. Carlson, J. Gordon, "Protoplasmic Viscosity Changes
     in Different Regions of the Grasshopper Neuroblast
     during Mitosis", Biol. Bull., 90, 109 (1946).   See
     also, Swanson, C. P. in Cytology and Cytogenetics,
     61, Figs. 3-10, Prentice Hall, New Jersey (1957).

32. Farnow, S. A. and Auld, B. A. "An Acoustic Phase Plate
     Imaging Device" presented at Sixth International
     Symposium on Acoustical Holography and Imaging,
     February 4-7, 1975, San Diego, California.

33. Kino, G. S. and Shaw, H. J. "Acoustic Surface Waves",
     Scientific American, 227, 51 (October 1972).

34. Havlice, J. F. et al. "An Electronically Focused
    Acoustic Imaging Device", in Acoustical Holography,
    5, 317, P. S. Green, Editor, Plenum Press, New York
    (1974).

35. Acoustical Holography, 5, P. S. Green, Editor, Plenum
    Press, New York (1974).

36. Photo, courtesy of W. P. Leung, H. J. Shaw and
    D. K. Winslow (unpublished).

37. Photo, courtesy of J. D. Fraser, T. M. Waugh and
    G. S. Kino (unpublished).

38. See Chapters 7 and 8 of this text.

Chapter 12

APPLICATIONS AND GENERAL CONCLUSIONS

C. F. Quate

Stanford University

Stanford, California 94305

The work presented in this series of lectures covers
a spectrum of ideas and techniques that are part of a
technology for imaging - a technology that has been expan-
ding and growing over the past several years. It was only
a short while ago that ultrasonic imaging was confined to a
few institutions and the instrumentation consisted of
variations of the system made up of a sound beam in the
form of a pencil-like probe that was scanned either by hand
or with the aid of mechanical levers. Many images were
recorded from a storage tube with two levels of brightness -
black and white. Research workers and clinicians have developed a great skill in the use of these instruments and
a vast amount of information has been recorded by a careful, systematic exploration and correlation with known
features. One prime example of the technique has been
described by Dr. F. Weill[1] and his co-authors in Besancon,
France.

One of the more important advances of rather recent
origin has been the introduction of grey scale into the
imaging by Dr. G. Kossoff[2] of Australia. This was done by
recording the image directly on film and the information
content was enormously improved - but, again, the instrumentation was not changed from the basic configuration.

We can learn from these lectures (and from the current
literature that surrounds this field) that the instrumentation is now changing in an important way. It is clear
that we will no longer be working with pencil beams of sound
as controlled in space by mechanical motion. It is clear

307

that we will soon have available focused beams, electronically controlled giving images in real time. In my view
this is of enormous importance. Why is this ? For several
reasons, real time imaging will allow us to use a maximum
range of grey scales and to use color coding to enhance our
ability to quickly interpret the images. An example of
this can be found from the work on geophysical exploration
for oil-bearing strata as illustrated on the back cover of
Science, 188, 6 June 1975.

But, of greater significance is the ability to view
motion with real time imaging. One of the most formidable
of problems encountered in ultrasonic imaging is that of
interpretation. The artifacts and the aberrations are
such that one is not able to use the direct interpretation
that is so familiar in optical photos. With relative
motion between two organs within the body cavity the task
of interpretation is much easier.

The coming technology with scanning can be roughly
divided into two distinct systems. In one system the
acoustic image is transferred to a deformable surface -
such as a liquid surface - and the deformations which
correspond to the pattern are scanned optically in a rapid
manner. In the second class of systems the pattern of
acoustic energy is read directly by an array of piezoelectric
elements. Each of these array systems has a special technique for reading and recording the amplitude and phase of
the acoustic signal at each point and they all contain the
piezoelectric array as an essential component in the system.
The systems which use optical scanning of liquid surfaces
have been farther advanced than the piezoelectric arrays,
but there is a fundamental problem that will one day limit
the growth of these systems.

Ultrasonic imaging will be used in those applications
where x-ray imaging can give rise to damage through radiation effects. The ultrasonic system does not suffer from
this disadvantage - provided that the power level is
sufficiently low. We know that acoustic power levels of
3 watts/cm^2 give rise to thermal effects which can be
damaging. It is likely that irreversible changes in cells
can occur at levels that are reduced by an order of
magnitude below this - but these levels are not firmly
established and those systems which operate with the lowest
power levels will surely have a strong competitive advantage

in the commercial world.   It may be this factor alone that
will determine the nature of the systems which eventually
evolve into instruments that find their way into widespread
use.

In a summary of this type one must refer to the
marvelous advances that have been made in x-ray tomography[3]
using reconstruction techniques for scanners used in
observing the head and with the "whole body scanners".
Some workers have observed that the clarity and resolution
in these new x-ray images are such that ultrasonic images
will not be needed.   This is a doubtful conclusion for as
is usual with advanced technology both techniques will be
used in combination and separately in specialized areas
within the body.

There are clear areas where ultrasound will be used.
In studies of the cardiac region such conditions as valve
abnormalities, vegetations and thrombus formation are
difficult to characterize and ultrasound will be exploited
for these problems.   The motion pattern is often characteristic of disease - the motion of the mitral valve is a
typical example.   One striking advantage of ultrasound is
the ability to see inside the aorta and observe blood clots.
We can, also, gain information on the elastic properties
and the compliance of the walls of these large vessels and
distinguish those which will not withstand the stress.
The study of the dynamics of flow through these vessels will
also be dependent upon the use of ultrasound.   The eye and
the orbit of the eye is still another region where ultrasound will be used extensively to diagnose malfunctioning
and imperfect regions.   In obstetrics ultrasound has a
clear advantage over x-rays and at present it is the area
where ultrasound is the most widely used.   This situation
will undoubtedly continue in the future since we know that
radiation of a normal fetus can increase the likelihood of
leukemia.   The growth of the fetus can be followed in
ultrasonic examinations and the fetal heart beat can be
easily monitored.   This represents another example of how
motion may be used to indicate trouble.   The normal beat
of 140 can increase to 160 or 170 with improper
oxygenation.

Such are examples of where it will be used.   But what
will be the growth pattern of this new art and what will be
the frequency of use ?

I, myself, have not worked a great deal in the field of ultrasonic imaging at the lower frequencies, but in preparing for this series of lectures and listening to the material that has been presented here, I have developed a feeling that we are now witnessing an upsurge in this field that will carry well beyond our present technology and permit clinical examination of much larger segments of our population.   One indication of this from data readily at hand comes from the overview article by Kossoff.[4] He reports that the number of patients examined with ultrasound at the National Acoustic Laboratory in Australia has increased from a level of 99 in the year 1964 to 1,380 in 1969 and to 4,238 in 1974.

We can foresee the advantages of using sound beams that are controlled by electronic means rather than with mechanical methods.   Different modes and different scanning techniques can be incorporated within the same instrument. The variety in the display will enhance the information content of the image.   But, there is still some way to go before we have a suitable technology for this form of imaging.

A major problem that is common to many systems of the present time relates to the artifacts that continually appear in those systems which rely on coherent or single frequency waves.   The diffraction effects from intervening and out-of-focus structures can be large enough to prevent interpretation of the images.   As Goodman[5] points out there are at least two pronounced difficulties with images derived with coherent illumination.   Firstly, tiny dust particles in the path of a plane wave will produce strong diffraction rings as the energy scattered from the particle interferes with the unscattered plane wave.   Secondly, the imaging properties of systems using coherent plane waves are inherently different from those that use incoherent waves. The difference is vividly illustrated by Considine.[6] He shows the response of a sharp edge for the two cases.   With the plane wave coherent illumination strong interference fringes appear in the vicinity of the edge - fringes which are completely absent if incoherent illumination is used.

There is an alternate method for eliminating the fringes. R. A. Lemons[7] has calculated the edge response for a focused beam wherein the edge was moved through the beam waist. It turns out in this case the edge response is very similar to the response obtained with incoherent energy. We thus conclude that with coherent systems we should avoid the use for plane waves for these will surely lead to extraneous diffraction patterns. Further, we should use focused systems whenever possible so as to obtain a proper response near the edges of the detail that make up the image.

It is more difficult to generate incoherent radiation in acoustic systems and it therefore has not been used in most systems. But Korpel[8] has demonstrated the advantages of incoherent ultrasonic imaging. He uses special trans-ducers for both the transmitting and receiving units and works with a wide spectrum of ultrasound that approximates the incoherent system. He has shown that this technique does reduce the interference fringes and it does aid in the interpretation of the images. But this interpretation will remain difficult until we learn much more about the ultrasonic characteristics of various tissues and organs within the body. This problem of tissue characterization will require some of our best effort if we are going to advance to the point where acoustic energy can be fully exploited in unraveling complex biological systems.

I am optimistic about the future and in part my optimism is based on the observation that there are other inhomogenous regions that have been explored with waves in much more detail and with greater results than has the human body. I am, of course, thinking of all the techniques that have been employed with electromagnetic waves. The atmosphere has been extensively studied and the inhomogeneities that occur there are in some ways better characterized than the response of tissue to ultrasonic waves. The use of x-rays in examining microscopic structures in crystalline and amorphous substances has been farther advanced than has the ultrasonic evaluation of tissue. As I look at the effort of individuals at various places around the country, I can see significant ideas coming from each endeavor. Allow me to enumerate on some of these. I won't dwell on the work that has been presented here. You have that well documented in the lectures and in the notes. But, I do want to mention some of the ideas that were not included in this set of lectures.

The work of Green at SRI has advanced the idea of electronically scanned piezoelectric arrays. The images of the human body using this system have been published in great detail.[9] This work is near the forefront of imaging with arrays and it serves to illustrate the advantages and the difficulties. One of the more remarkable advantages is the near real time imaging (3 frames per second) which allows one to see motion. As mentioned previously, this is of enormous help in interpreting the images.

The work of the group with Thurstone[10] at Duke University is significant for they have shown how to use a computer to assimilate and process both the amplitude and phase of the signal picked up from each element of an array. The computer control can be used to alter the phase in a way that simulates the action of a lens with various focal lengths. At present the computer cost dominates the cost of the entire instrument but with special purpose micro-processors this cost could in principle be reduced.

The work of Greenleaf[11] and his colleagues at the Mayo Foundation has provided us with background information on how the velocity profile of tissues can be used to record tomographic images. Their work should allow us to use "reconstruction techniques"[3] for imaging various internal organs.

The work of Mezrich at RCA Laboratories has shown[12] that the attenuation characteristics for sound near 1.5 MHz in excised breast tissue is markedly dependent on the pathology of the tissue. That group has found that normal tissue exhibits an attenuation of 1.5 dB/cm whereas tissue with a benign tumor exhibits an attenuation near 6 dB/cm. And, still more interesting is the fact that in tissue with malignant tumors the attenuation increases to 14 dB/cm !!

Dr. J. Birnholz at Stanford in unpublished work has observed that the attenuation of sound at 2.5 MHz in certain soft tissues is critically dependent upon the temperature of the tissue. A change of only a few degrees can make an observable difference in the sound attenuation for a path length of a few centimeters. This could be important for there are temperature differences within the body that could in principle be detected with this technique. It should also serve as a caution in using the attenuation data from excised tissue since the temperature may be different from the _in vivo_ tissue.

Mezrich[13] has also demonstrated that phase contrast
techniques can be incorporated into a sound system in a
manner almost identical to that of the optical system.
He has shown the marked advantage of phase contrast in the
imaging of those features which do not alter the wave
amplitude but do have a change in velocity.

As a final point we should include the excellent work
that has been done by Waag and Gramiak[14] at Rochester.
As mentioned earlier, electromagnetic radiation is often
used to characterize the spatial distributions of inhomogeneities in the index of refraction within a given region
to be explored.   What is required in most instances is a
wide band of frequencies.   The spatial information is
obtained by studying the Fourier spectrum of that scattered
radiation.   The same ideas can be carried over into the
field of ultrasonics.   In their preliminary work the
Rochester group has demonstrated that the frequency spectrum
of the scattered radiation can be markedly changed in tissue
showing pathology when compared to normal tissue.   It is
their hope that the frequency spectrum can be used as a
"signature" to identify various diseases.   We wish them
luck.

With all of this before us what can we say in summary.
Numerous ideas exist in a variety of laboratories and yet
there seems to be no central focus where it can all come
together in the form of a clinical instrument that can be
widely distributed to medical centers throughout the
country.

What is needed is an institute at the national level -
an institute that can furnish laboratory facilities and
funding where the spectrum of ideas that exist throughout
the country can be combined and sifted and used in the
development of a technology that is available to all
clinical workers.   The National Acoustic Institute of
Australia as directed by Dr. G. Kossoff can be used as the
model.[4]   They have managed to fund and operate an institute
with a range of activities from research to the clinical
examinations of patients all within the confines of a single
unit.   Their technology and knowledge represents a leading
effort as measured on a world-wide scale.   The work is
innovative and some of the most accurate and most revealing

images have come from this Institute.   And yet even there
a large number of the ideas outlined above have not been
exploited.

The features of the new technology - the ability to
scan in real time in either the B-mode or the C-mode - the
ability to observe and record motion - to observe the "way
things move" - the ability to exploit temperature
variations and variations in attenuation and the ability to
incorporate a wide spectrum of sound frequencies should
improve the speed of making accurate diagnosis in a fundamental way.

It was Kossoff who wrote[4] ". . . the present clinical
acceptance of ultrasonic diagnostic techniques is based on
the utilization of one acoustic parameter of the tissue -
i.e. the acoustic impedance mismatch".   The other acoustic
parameters referred to above are not yet part of the
clinical technique.   These new features when combined with
the recent advances that have been achieved with x-ray
tomography should allow one to "see" inside the body with
a clarity that was hard to contemplate a few short years
ago.   In discussing advances in the field of oceanography
someone has suggested that we adjust our research program
so that one day the "sea will become transparent".   I am
convinced that the technology of imaging will soon be
advanced to the point where "the human body will become
transparent".

REFERENCES

1. Weill, F. et al. "Ultrasonic Study of Venous Patterns
    in the Right Hypochrondium: An Anatomical Approach to
    Differential Diagnosis of Obstructive Jaundice",
    J. Clinical Ultrasound, 3, 23 (March 1975).   Errata 3,
    153 (June 1975).

2. Kossoff, G., Garrett, W. J. and Radovanovich, G. "Grey
    Scale Echography in Obstetrics and Gynecology",
    Australian Radiology, 68, 62 (March 1974).

3. Hounsfeld, G. N. "Computerized Transverse Axial
    Scanning (Tomography): Part 1 - Description of the
    System", British J. Radiology, 46, 1016 (1973).   See
    also, Ambrose, J. "Computerized Transverse Axial

Scanning (Tomography): Part 2 - Clinical Application",
British J. Radiology, 46, 1023 (1973).

4. Kossoff, G. "An Historical Review of Ultrasonic
    Investigations at the National Acoustic Laboratories",
    J. Clinical Ultrasound, 3, 39 (March 1975).

5. Goodman, J. W. Introduction to Fourier Optics,
    Section 6-5, 125, McGraw-Hill Book Co., New York (1968).

6. Considine, P. S. "Effects of Coherence on Imaging
    Systems" J. Optical Soc. Am., 56, 1001 (1966).

7. Lemons, R. A. "Acoustic Microscopy by Mechanical
    Scanning", Internal Memorandum, 74 (May 1975).

8. Korpel, A., Whitman, R. L. and Ahmed, M. "Elimination
    of Spurious Detail in Acoustic Images", Acoustical
    Holography, 5, 373, P. S. Green, Editor, Plenum Press,
    New York (1974).

9. Marich, K. W. et al. "Real Time Imaging with a New
    Ultrasonic Camera: Part 1, In Vitro Experimental
    Studies on Transmission Imaging of Biological
    Structures" J. Clinical Ultrasound, 3, 5 (March 1975).

10. Thurstone, F. L. and von Ramm, O. T. "A New Ultrasound
    Imaging Technique Employing Two-Dimensional Electronic
    Beam Steering", Acoustical Holography, 5, 249,
    P. S. Green, Editor, Plenum Press, New York (1974).

11. Greenleaf, J. F. et al. "Algebraic Reconstruction of
    Spatial Distributions of Acoustic Velocities in Tissue
    from their Time of Flight Profiles", Acoustical
    Holography, 6, Plenum Press, New York (in press).

12. Mezrich, R. S., Etzold, K. F. and Vilkomerson, D.H.R.
    "Ultrasonovision" 1974 Ultrasonics Symposium
    Proceedings, IEEE Cat. #74CHO869-1SU, 1 (1974).

13. Mezrich, R. S. "Phase Contrast in Ultrasonovision"
    (to be published).

14. Waag, R. C. and Lerner, R. M. "Tissue Macrostructure
    Determination with Swept Frequency Ultrasound" 1973
    Ultrasonics Symposium Proceedings, IEEE Cat. #73CHO-
    807-8SU, 63 (1973).